SpringerBriefs in Electrical and Computer Engineering

Computational Electromagnetics

Series editor

Rakesh Mohan Jha, Bangalore, India

More information about this series at http://www.springer.com/series/13885

Balamati Choudhury · Rakesh Mohan Jha

Refined Ray Tracing Inside Single- and Double-Curvatured Concave Surfaces

Springer

Balamati Choudhury
Centre for Electromagnetics
CSIR-National Aerospace Laboratories
Bangalore, Karnataka
India

Rakesh Mohan Jha
Centre for Electromagnetics
CSIR-National Aerospace Laboratories
Bangalore, Karnataka
India

ISSN 2191-8112   ISSN 2191-8120   (electronic)
SpringerBriefs in Electrical and Computer Engineering
ISSN 2365-6239   ISSN 2365-6247   (electronic)
SpringerBriefs in Computational Electromagnetics
ISBN 978-981-287-807-6   ISBN 978-981-287-808-3   (eBook)
DOI 10.1007/978-981-287-808-3

Library of Congress Control Number: 2015947813

Springer Singapore Heidelberg New York Dordrecht London

Springer Science+Business Media Singapore Pte Ltd. is part of Springer Science+Business Media
(www.springer.com)

To Professor Pravas R. Mahapatra

In Memory of Dr. Rakesh Mohan Jha
Great scientist, mentor, and excellent
human being

Dr. Rakesh Mohan Jha was a brilliant contributor to science, a wonderful human being, and a great mentor and friend to all of us associated with this book. With a heavy heart we mourn his sudden and untimely demise and dedicate this book to his memory.

Preface

Most of the complex aerospace structures may be modeled as hybrid of quadric cylinders (QUACYLs) and quadric surfaces of revolution (QUASORs). Although numerically specified surfaces (NURBS) too are used for modeling of aerospace surfaces, for most practical EM applications, it is sufficient to model them as quadric surface patches and the hybrids thereof for geometric ray tracing. Geometric ray tracing is a priori requirement for analyzing the RF build-up inside these electrically large aerospace bodies. Toward this, ray tracing inside different quadric surfaces such as right circular cylinder, GPOR, GPOR frustum of different shaping parameters have been carried out, and the corresponding ray path are visualized. Finally, ray tracing inside a space module, which is a hybrid of a finite segment of right circular cylinder and a frustum of general paraboloid of revolution (GPOR), is analyzed for practical aerospace applications.

Balamati Choudhury
Rakesh Mohan Jha

Acknowledgments

We would like to thank Mr. Shyam Chetty, Director, CSIR-National Aerospace Laboratories, Bangalore for his permission and support to write this SpringerBrief.

We would also like to acknowledge valuable suggestions from our colleagues at the Centre for Electromagnetics, Dr. R.U. Nair, Dr. Hema Singh, Dr. Shiv Narayan, and Mr. K.S. Venu during the course of writing this book.

We wish to recognize the contributions of Ms. Gouramma Hiremath and Ms. Sanjana Bisoyi in this area, during their training at the Centre for Electromagnetics, CSIR-NAL.

But for the concerted support and encouragement from Springer, especially the efforts of Suvira Srivastav, Associate Director, and Swati Meherishi, Senior Editor, Applied Sciences & Engineering, it would not have been possible to bring out this book within such a short span of time. We very much appreciate the continued support by Ms. Kamiya Khatter and Ms. Aparajita Singh of Springer toward bringing out this brief.

Contents

About the Authors

Dr. Balamati Choudhury is currently working as a scientist at Centre for Electromagnetics of CSIR-National Aerospace Laboratories, Bangalore, India since April 2008. She obtained her M.Tech. (ECE) degree in 2006 and Ph.D. (Engg.) degree in Microwave Engineering from Biju Patnaik University of Technology (BPUT), Rourkela, Orissa, India in 2013. During the period of 2006–2008, she was a Senior Lecturer in the Department of Electronics and Communication at NIST, Orissa India. Her active areas of research interests are in the domain of soft computing techniques in electromagnetics, computational electromagnetics for aerospace applications, and metamaterial design applications. She was also the recipient of the CSIR-NAL Young Scientist Award for the year 2013–2014 for her contribution in the area of Computational Electromagnetics for Aerospace Applications. She has authored and co-authored over 100 scientific research papers and technical reports including a book and three book chapters. Dr. Balamati is also an Assistant Professor of AcSIR, New Delhi.

Dr. Rakesh Mohan Jha was Chief Scientist & Head, Centre for Electromagnetics, CSIR-National Aerospace Laboratories, Bangalore. Dr. Jha obtained a dual degree in BE (Hons.) EEE and M.Sc. (Hons.) Physics from BITS, Pilani (Raj.) India, in 1982. He obtained his Ph.D. (Engg.) degree from Department of Aerospace Engineering of Indian Institute of Science, Bangalore in 1989, in the area of computational electromagnetics for aerospace applications. Dr. Jha was a SERC (UK) Visiting Post-Doctoral Research Fellow at University of Oxford, Department of Engineering Science in 1991. He worked as an Alexander von Humboldt Fellow at the Institute for High-Frequency Techniques and Electronics of the University of Karlsruhe, Germany (1992–1993, 1997). He was awarded the Sir C.V. Raman Award for Aerospace Engineering for the Year 1999. Dr. Jha was elected Fellow of INAE in 2010, for his contributions to the EM Applications to Aerospace Engineering. He was also the Fellow of IETE and Distinguished Fellow of ICCES. Dr. Jha has authored or co-authored several books, and more than five hundred scientific research papers and technical reports. He passed away during the production of this book of a cardiac arrest.

Abbreviations

EM   Electromagnetics
GPOR   General paraboloid of revolution
QUACYLs   Quadric cylinders
QUASORs   Quadric surfaces of revolution

List of Figures

List of Tables

Refined Ray Tracing Inside Single- and Double-Curvatured Concave Surfaces

Abstract Analytical surface modeling is a priori requirement for electromagnetic (EM) analysis over aerospace platforms. Although numerically specified surfaces and even non-uniform rational basis spline (NURBS) can be used for modeling such surfaces, for most practical EM applications, it is sufficient to model them as quadric surface patches and the hybrids thereof. It is therefore apparent that a vast majority of aerospace bodies can be conveniently modeled as combinations of simpler quadric surfaces, i.e., hybrid of quadric cylinders and quadric surfaces of revolutions. Hence the analysis of geometric ray tracing inside is prerequisite to analyzing the RF build-up. This book describes the ray tracing effects inside different quadric surfaces such as right circular cylinder, general paraboloid of revolution (GPOR), GPOR frustum of different shaping parameters, and the corresponding visualization of the ray-path details. Finally ray tracing inside a typical space module, which is a hybrid of a finite segment of right circular cylinder and a frustum of GPOR is analyzed for practical aerospace applications.

Keywords Refined ray tracing · Quadric cylinders · Quadric surface of revolution · Space module · Aerospace structures

1 Introduction

Analytical surface modeling is a priori requirement for electromagnetic (EM) analysis over aerospace platforms. Although numerically specified surfaces and even NURBS can be used for modeling such surfaces, for most practical EM applications, it is sufficient to model them as quadric surface patches and the hybrids thereof. It is therefore apparent that a vast majority of aerospace bodies can be conveniently modeled as combinations of simpler quadric surfaces. The most common body shape is cylindrical, with a conical structure in the front. The quadrics most commonly used as aerospace shapes are the quadric cylinders and surfaces of revolution. The quadric cylinders (QUACYLs) consist of circular,

© The Author(s) 2016
B. Choudhury and R.M. Jha, *Refined Ray Tracing Inside Single-
and Double-Curvatured Concave Surfaces*, SpringerBriefs
in Computational Electromagnetics, DOI 10.1007/978-981-287-808-3_1

elliptic, hyperbolic, and parabolic cylinders. Similarly, the right circular cone, sphere, ellipsoid, hyperboloid, and paraboloid of revolution constitute the quadric surfaces of revolution (QUASORs) (Jha and Wiesbeck 1995). These QUACYLs and QUASORs are the coordinate surfaces from the eleven Eisenhart Coordinate Systems, and are easily generated by keeping one of the orthogonal coordinate parameters constant (Moon and Spencer 1971). One advantage of parameterizing the surfaces is the availability of *shaping parameters* (Jha and Wiesbeck 1995), in terms of the coordinate held constant.

In this work, the effect of shaping parameter on the ray tracing over single- and double-curvatured surfaces is reported. Ray tracing inside different quadric surfaces (such as right circular cylinder, general paraboloid of revolution (GPOR), GPOR frustum, and a hybrid of GPOR and right circular cylinder) are carried out, and the ray path details are visualized (Fig. 1). Finally, the ray tracing inside a space module, which is a hybrid of a finite segment of right circular cylinder and a frustum of GPOR, is reported. Section 2 of this book gives the analytical surface modeling of the above-mentioned shapes. Section 3 describes the developed refined ray tracing algorithm along with the visual explanation and Sect. 4 discusses the geometric ray tracing effect inside the single- and double-curvatured surfaces.

2   Modeling of Quadric Surfaces and Hybrids

Unlike the ray tracing packages, a quasi-analytical ray tracing method is implemented here in conjunction with analytical surface modeling of the quadric surfaces. This section describes the analytical expression to generate the quadric surfaces along with their surface visualization.

2.1   Right Circular Cylinder

A right circular cylinder can be used for first-order modeling of the fuselage of the space module. The parametric equations of right circular cylinder are given as (Jha and Wiesbeck 1995):

$$x = a \cos u; y = a \sin u; z = v1 : v2 \tag{1}$$

where a is the radius of the right circular cylinder, which varies from zero to infinity, u is the angle, varying from 0° to 360°, and v is the height of the right circular cylinder. A right circular cylinder of length 16.5 m and diameter 4.5 m is considered (Fig. 2) and modeled using *Matlab* (Fig. 3).

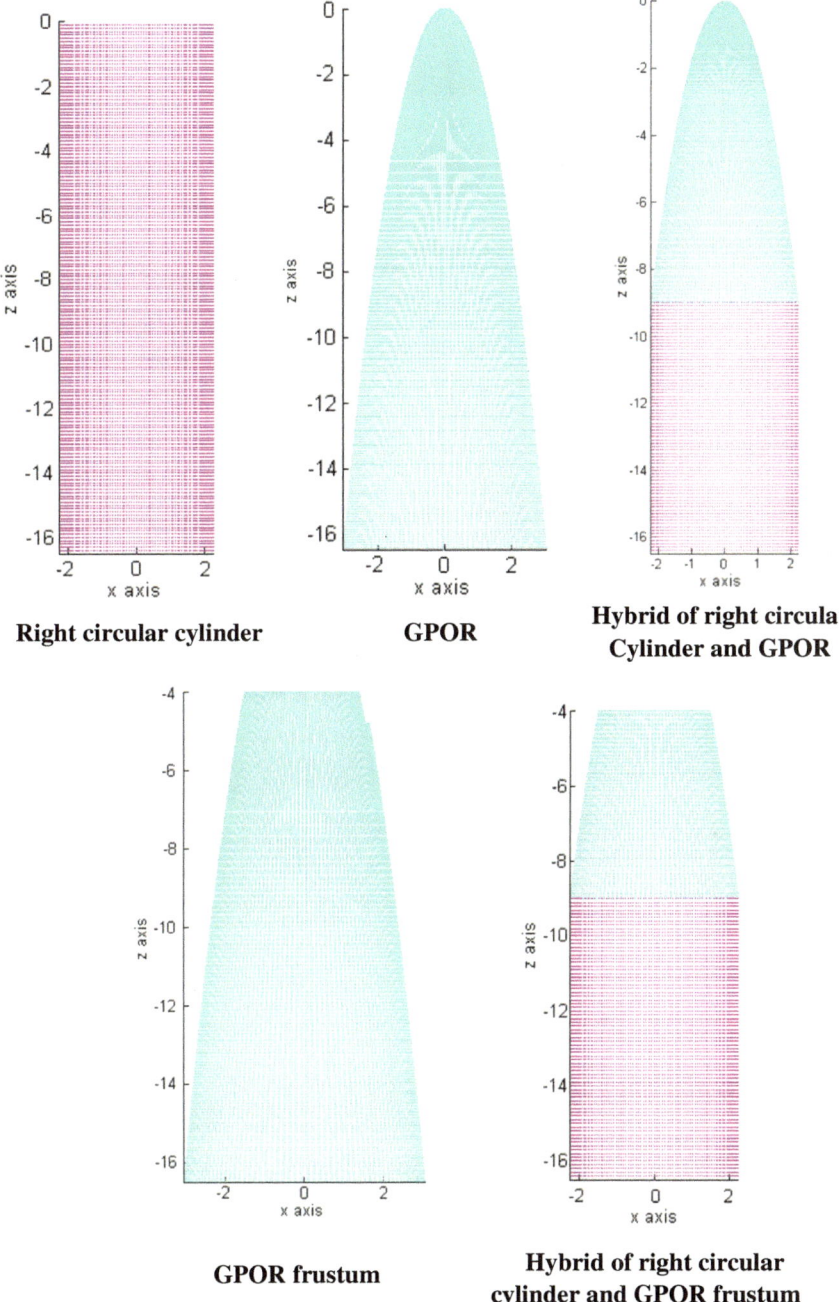

Right circular cylinder   **GPOR**   **Hybrid of right circular Cylinder and GPOR**

GPOR frustum   **Hybrid of right circular cylinder and GPOR frustum**

Fig. 1 Quadratics and their hybrids considered for analysis of ray propagation mechanism

Fig. 2 Dimensions of a
typical right circular cylinder

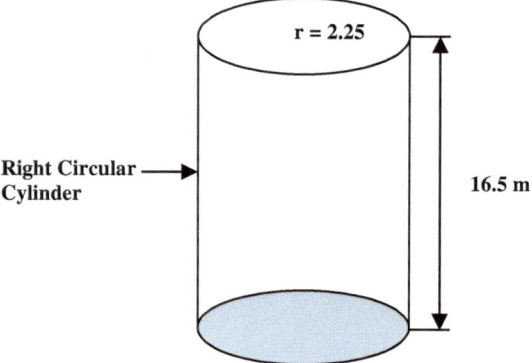

Fig. 3 The right circular
cylinder modeled using
Matlab

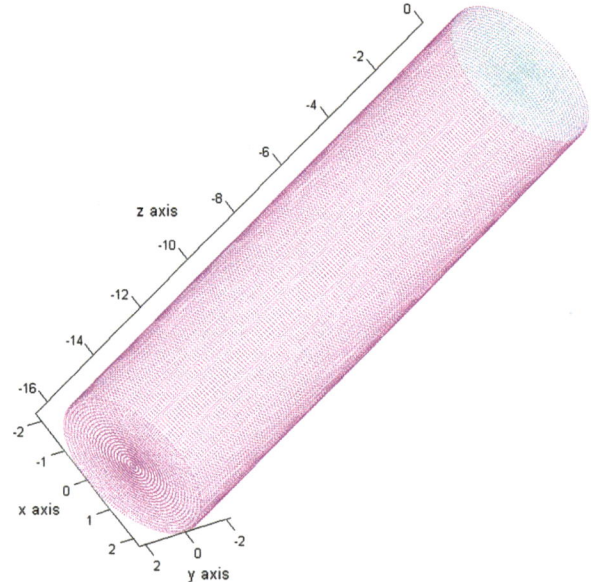

2.2 *General Paraboloid of Revolution (GPOR)*

A general paraboloid of revolution can be generated by rotating a parabola in a particular axis. This surface can model the nose cone of an aircraft. The parametric equations of paraboloid are given as (Jha and Wiesbeck 1995):

$$x = au \cos v, \; y = au \sin v, z = -u^2 \tag{2}$$

where a is the distance between the vertex and focus, which varies from zero to infinity, v is the angle varying from 0° to 360°, and u is the height. A GPOR of

Fig. 4 Dimensions of a
typical GPOR

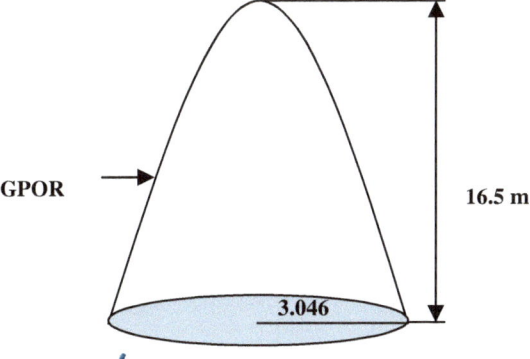

Fig. 5 The GPOR modeled
using *Matlab*

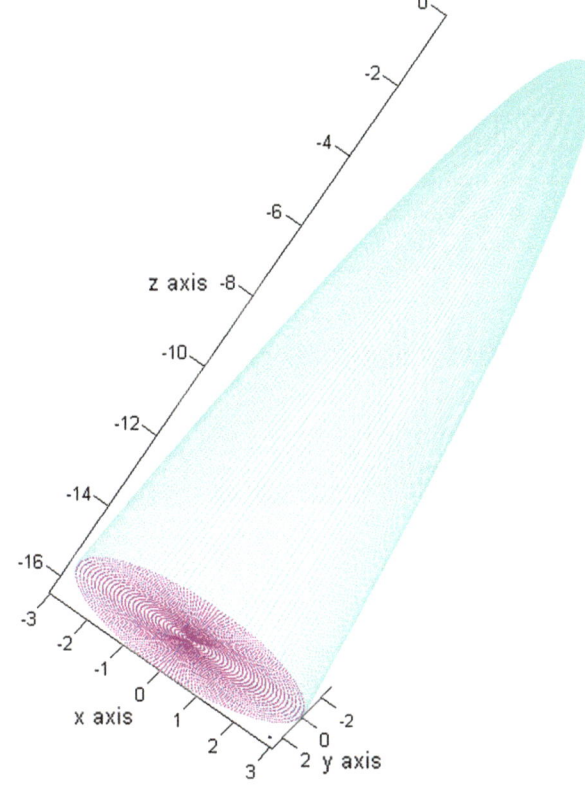

length (height) 16.5 m is shown in Fig. 4. The shaping parameter for the GPOR is
assumed to be 0.75. The lower radius r of GPOR is given by $r = au = 3.0465$ m.
The modeled GPOR is visualized using *Matlab* (Fig. 5).

Fig. 6 Dimensions of a
typical GPOR frustum

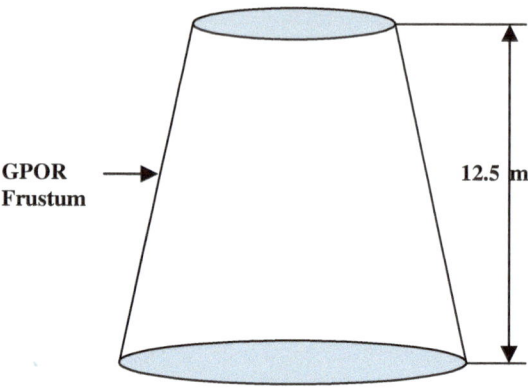

2.3 GPOR Frustum

GPOR frustum is a part of GPOR and it can be modeled as a part of space module.
For the GPOR frustum, shaping parameter is assumed to be 0.75 such that the basis
parameter coordinate varies from $u_1 = 2$ to $u_2 = 3$. A matching lower radius of
2.25 m results in the required upper radius of the GPOR frustum of 1.5 m (Fig. 6).
The modeled GPOR frustum is visualized using *Matlab* (Fig. 7).

2.4 Hybrid of GPOR and Right Circular Cylinder

This structure can be modeled as a hybrid of a general paraboloid of revolution
(GPOR) and a finite segment of right circular cylinder. The internal dimensions
(Fig. 8) of a typical hybrid structure considered are given as follows (Jackson 2007):

Right circular cylinder length: 7.5 m;
GPOR frustum height: 9 m; and
Right circular cylinder diameter: 4.5 m.

The shaping parameter for the GPOR frustum is assumed to be 0.75. Modeling
and visualization of the hybrid structure using *Matlab* is shown in Fig. 9.

2.5 Manned Space Module

A manned space module can be modeled as a hybrid of a general paraboloid of
revolution (GPOR) frustum and a finite segment of right circular cylinder. The
internal dimensions of a typical manned space module considered (Jackson 2007)
are given as follows:

Fig. 7 The GPOR frustum
modeled using *Matlab*

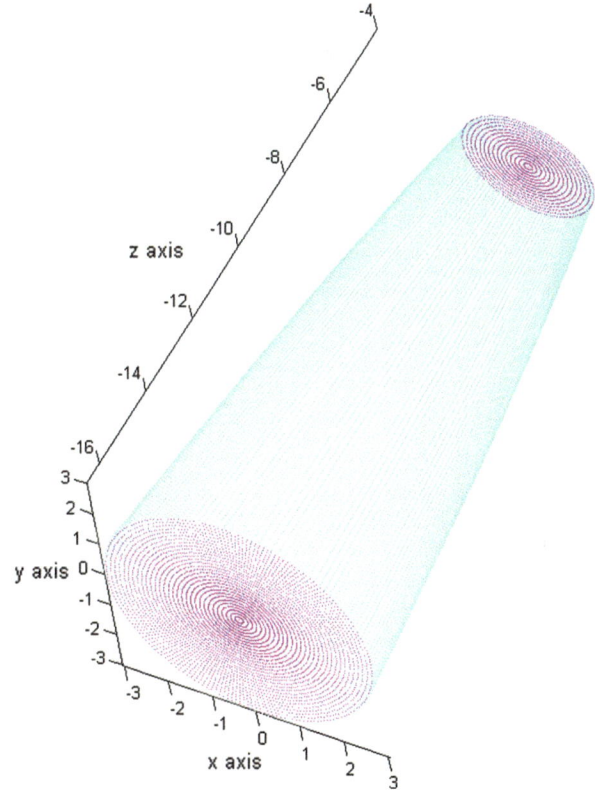

Fig. 8 Dimensions of a
typical GPOR–right circular
cylinder

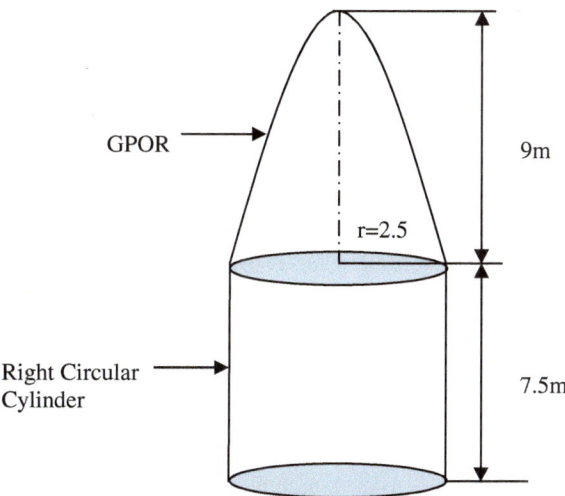

Fig. 9 The GPOR–right
circular cylinder modeled
using *Matlab*

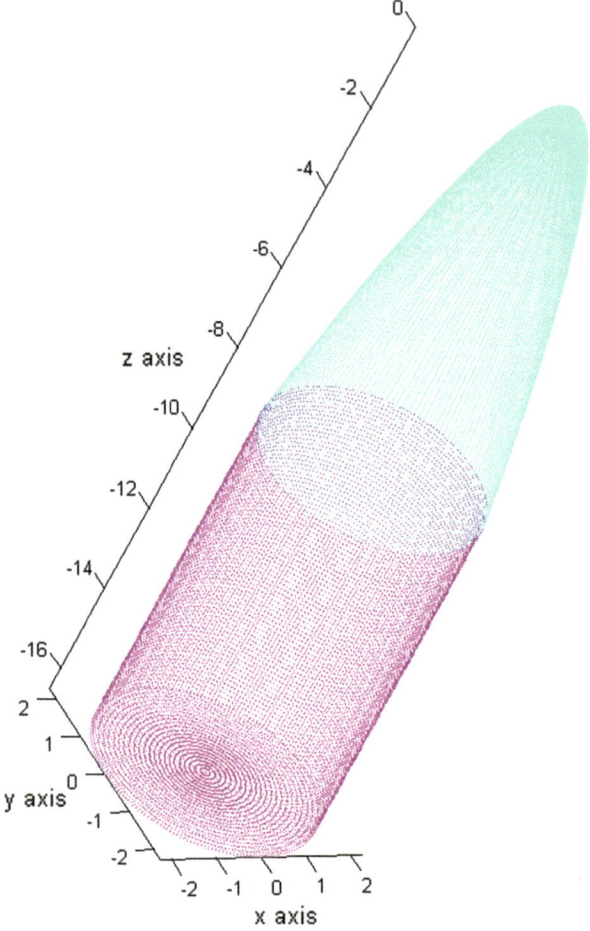

Right circular cylinder length: 7.5 m;
GPOR frustum height: 5 m; and
Right circular cylinder diameter: 4.5 m.

For the GPOR frustum, shaping parameter is assumed to be 0.75 such that the basis
parameter coordinate varies from $u_1 = 2$ to $u_2 = 3$. A matching lower radius of 2.25 m
results in the required upper radius of the GPOR frustum of 1.5 m (Figs. 10 and 11).

3   Refined Ray Tracing Technique

In refined ray tracing technique, a transmitter and a receiver are placed inside the
cavity and the rays are launched from the transmitter using a *uniform ray launching
scheme* (Seidel and Rappaport Seidel and Rappaport 1994). Each ray is defined by

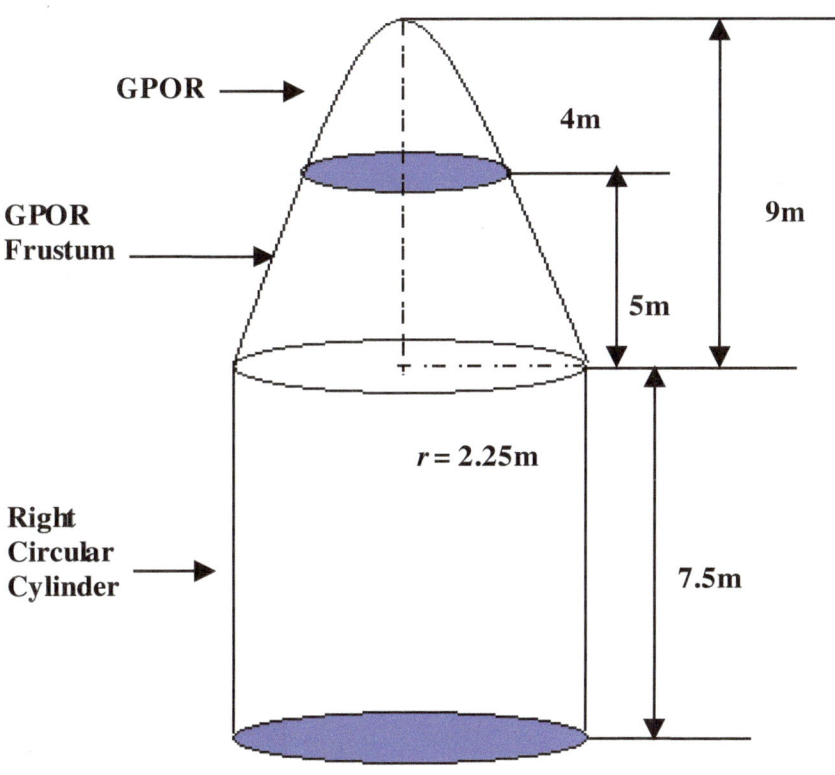

GPOR ⟶

GPOR
Frustum ⟶

Right
Circular
Cylinder ⟶

4m

9m

5m

$r = 2.25$m

7.5m

Fig. 10 Dimensions of a typical manned space module

their (θ, ϕ) values. The rays are then allowed to propagate inside the cabin. The first intersection point is determined using intersection formula between the line and the corresponding surface equation (Kreyszig 2010). According to the z-coordinate of the first intersection point, the first incident point is calculated using the respective surface equations. The normal at the first incident point is determined by taking the surface normal equation of the corresponding surfaces (Jha and Wiesbeck 1995). The intermediate point on the reflected ray is then obtained using the Snell's law of reflection, i.e., by taking the angle of incidence equal to the angle of reflection and imposing the condition of co-planarity. The same process is repeated for the given number of bounces.

Fig. 11 The space module
modeled using *Matlab*

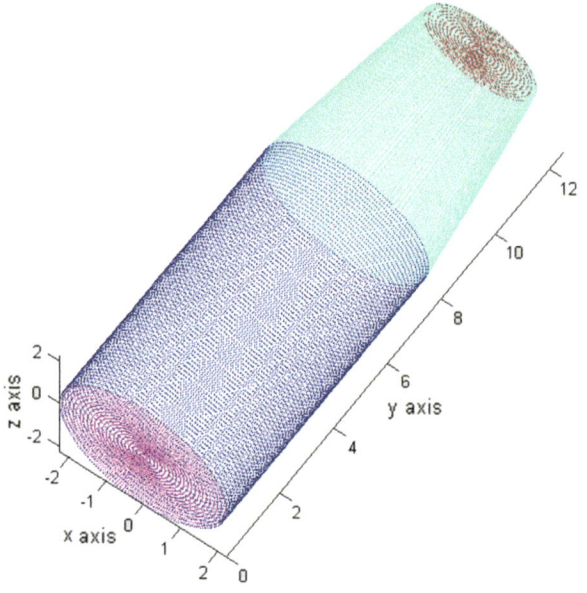

Fig. 12 Bunches of rays
shown with a unit isotropic
source that reach the receiver

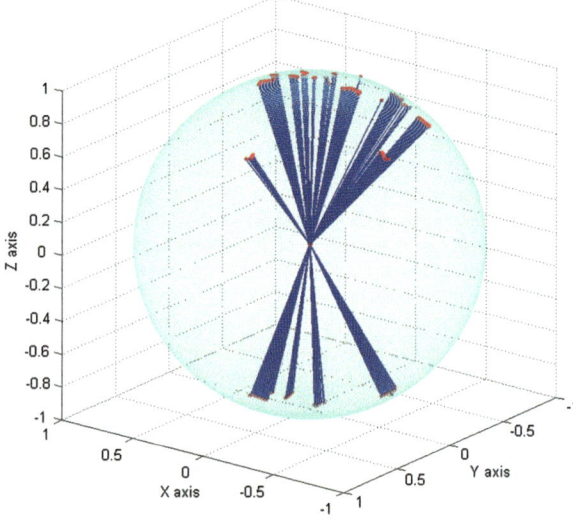

The receiver is considered as a small sub-cube placed inside the curvatured structures. The center of the sub-cube is the observation point.

The rays that reach the reception sub-cube are considered as the rays required for the field build-up inside the cabin. The launched rays from an isotropic source are shown in Fig. 12. The ray paths tend to appear as *ray bunches* (Fig. 13), which traverse in a nearly parallel manner and reach the receiver. Hence, an algorithm is developed for identifying these bunches. The *ray within a bunch*, which is closest to the receiving point, is taken as the test ray for convergence (Red color ray: Fig. 14).

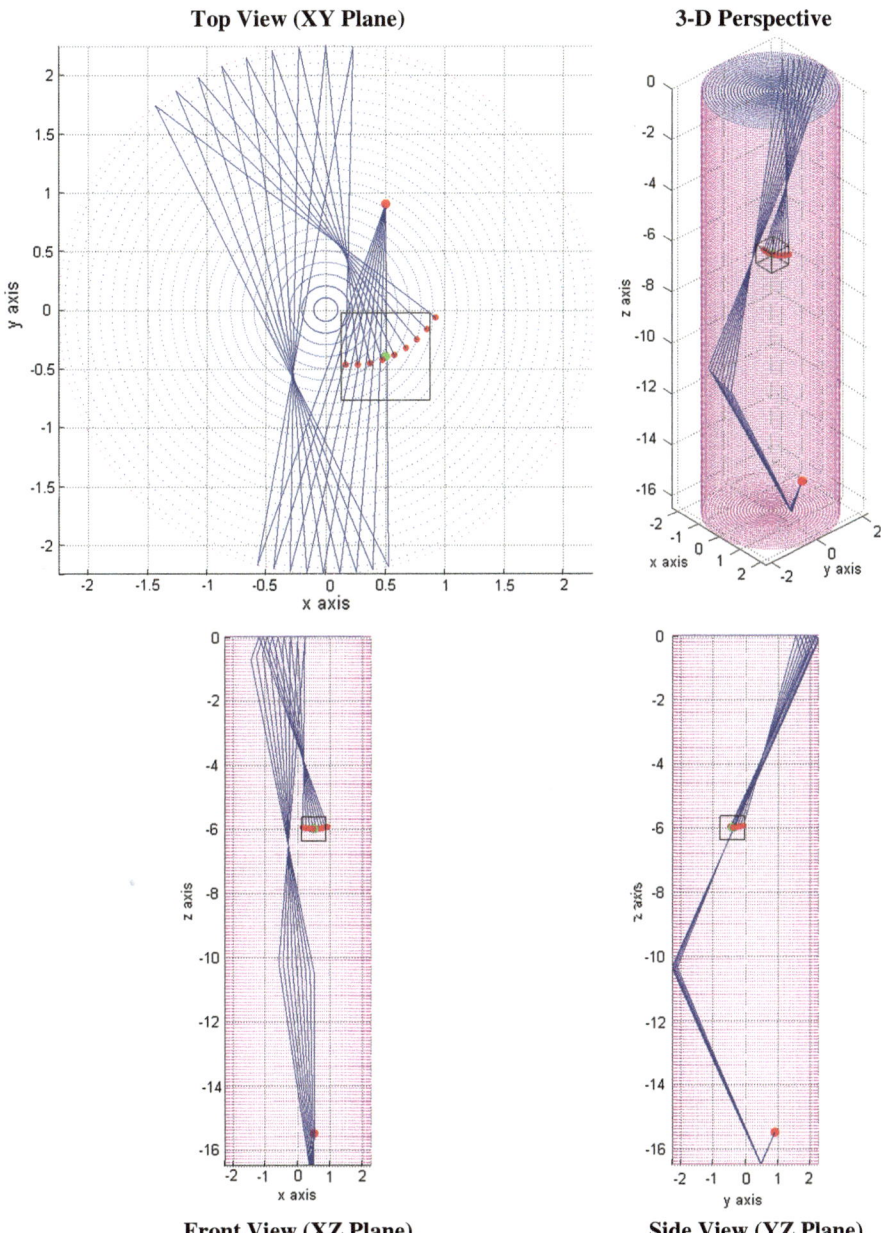

Fig. 13 A single bunch of rays reaching the receiver. *Green dot* represents the center of the reception sub-cube

This ray is converged by refining the angular separation iteratively (Green color ray: Fig. 15), to yield the ray solutions at the receiver after refinement. Then the ray path details of the identified ray are written in a file.

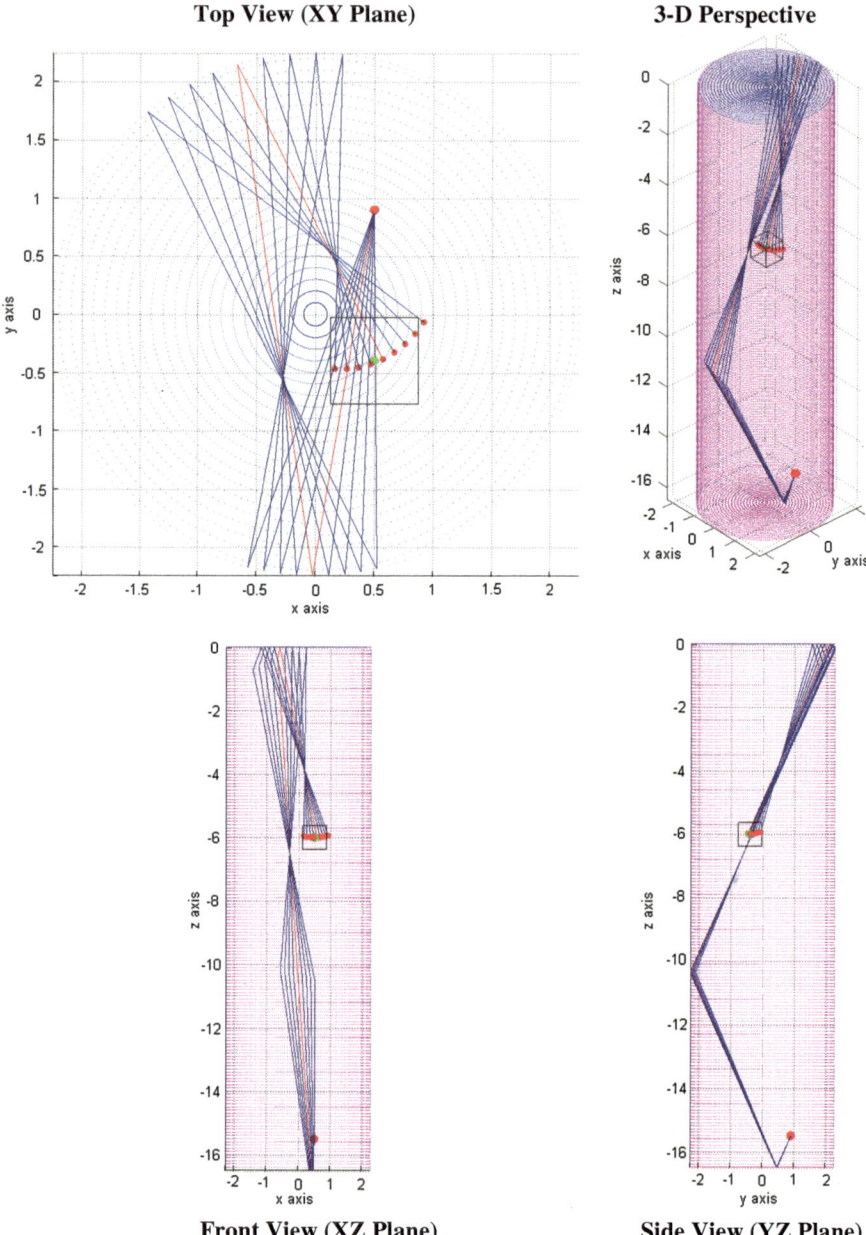

Fig. 14 A single ray that is closest to the receiving point is selected from the bunch as shown in Fig. 13 (*Red color* ray is the representative ray chosen. *Green dot* represents the center of the reception sub-cube)

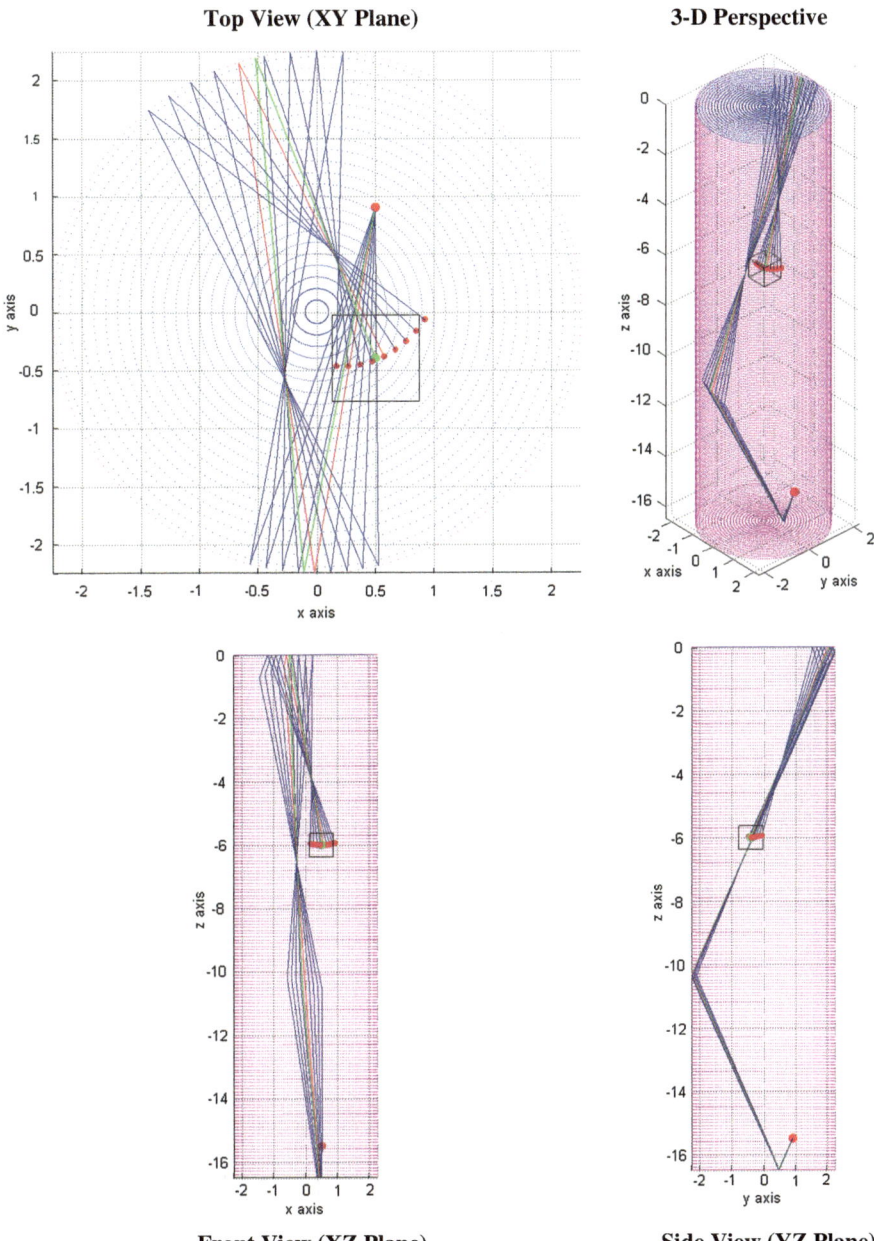

Fig. 15 The selected ray (*red color*) from the ray bunch is refined to converge the reception point (*Green color* ray is the refined ray. *Green dot* represents the center of the reception sub-cube)

4 Ray Propagation Inside the Quadric Surfaces and Hybrids

The refined ray tracing algorithm is implemented inside the above-mentioned quadric surfaces. The transmitter and receiver are placed inside the quadric surfaces at S(0.5, 0.9, −15.5) and R(0.5, −0.4, −6.0) such that the ray tracing effect w.r.t. different quadratic surfaces can be analyzed.

4.1 Ray Tracing Inside the Right Circular Cylinder

After modeling the right circular cylinder, a transmitter and a receiver are placed randomly inside the right circular cylinder. As explained in refined ray tracing technique in Sect. 3, the first intersection point is calculated using intersection formula between the line and the surface of right circular cylinder. As the structure has three surfaces at different heights, the z-coordinate of the first intersection point is checked and the equation is adopted for calculation of *first incident point* according to the surface. The corresponding three surfaces are given below:

(a) At $z = 0$, surface is a plane;
(b) If $0 \geq z > -16.5$, surface is a right circular cylinder; and
(c) If $z = -16.5$, surface is a plane.

Then the normal at the first incident point is determined by taking the normal equation of the corresponding surfaces (Jha and Wiesbeck 1995).

The normal at the right circular cylinder is given by

$$x_n = a \cos \phi \tag{3.a}$$

$$y_n = a \sin \phi \tag{3.b}$$

$$z_n = 0 \tag{3.c}$$

where a is the radius of the right circular cylinder. The ray path details are calculated using the refined ray tracing algorithm.

The surface normal equation of the right circular cylinder, described in Sect. 4.1, is used and visualized (Fig. 16). This normal line is used to check the co-planarity condition and law of equal angles.

4.1.1 Results and Visualization of Ray Path

The ray path propagation inside the right circular cylinder is visualized using *Matlab*. Table 1 shows that the 24 rays reach the receiver cumulatively up to three bounces (excluding the direct ray). There are four rays, which reach after one

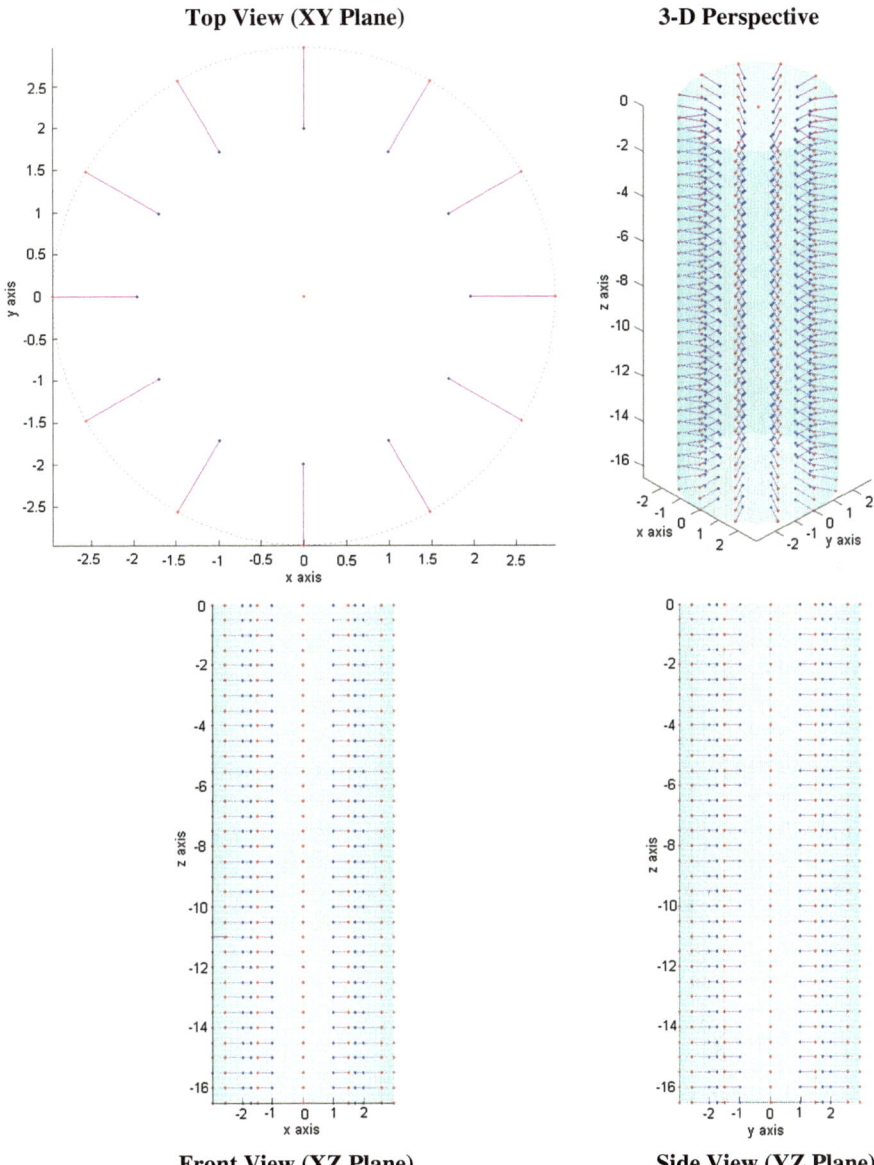

Fig. 16 Surface normal visualization of right circular cylinder at different incident points

bounce, eight rays after two bounces, and 12 rays after three bounces. Figures 17, 18, and 19 give the visualization of one-bounce, two-bounce, and three-bounce rays reaching receiver. Figure 20 gives the cumulative number of rays till the third bounce.

Table 1 Description of 24 rays that reached the receiver (till 3 bounce) out of 4,126,183 rays launched at 0.1° angular separation

Sl. no	θ	φ	Ex	Ey	Ez	Path length (m)	Time (μs)
1	40.173	97.426	−1.944	−1.325	−2.787	19.145	0.063
2	31.151	213.503	−2.595	−0.418	−4.349	19.604	0.065
3	47.709	258.964	−1.020	2.588	−2.531	20.140	0.067
4	53.447	37.180	−0.512	2.649	−2.001	21.457	0.071
5	145.106	97.426	−1.944	−1.325	−3.373	21.590	0.071
6	137.755	258.964	−1.020	2.588	−3.064	22.160	0.073
7	20.515	22.613	1.246	1.819	−5.892	22.455	0.074
8	153.464	213.503	−2.595	−0.418	−5.265	22.700	0.075
9	57.447	222.383	−2.749	0.380	−1.772	22.937	0.076
10	162.823	22.613	1.246	1.819	−7.133	25.783	0.085
11	173.550	269.999	4.24E−06	1.186	−10.5	28.178	0.093
12	25.909	258.964	−1.020	2.589	5.729	34.097	0.113
13	20.457	97.3869	−1.849	−1.261	6.000	35.009	0.116
14	14.953	213.504	−1.582	−0.255	6.000	35.322	0.117
15	166.268	213.503	−1.447	−0.233	5.999	37.821	0.126
16	3.460	270.003	−2.3E−05	0.362	5.999	38.069	0.126
17	9.389	22.557	0.560	0.818	5.999	38.515	0.128
18	176.833	270	1.59E−08	0.331	6	40.061	0.133
19	171.397	22.556	0.513	0.748	5.999	40.455	0.134
20	1.750	269.976	0.000	0.320	−10.5	59.027	0.196
21	4.799	22.734	0.497	0.727	−10.498	59.206	0.197
22	7.697	213.580	−1.400	−0.229	−10.499	59.536	0.198
23	178.351	269.982	9.16E−05	0.302	−10.500	61.025	0.203
24	1.373	269.972	7.01E−05	0.143	5.999	71.020	0.236

4.1.2 Discussion

A convergence study of the number of rays that reach the receiving point w.r.t. the number of rays launched (a function of angular separation) is carried out. The cumulative rays (till 10 bounces) that reach the receiver after refinement are given in Table 2. A close scrutiny on Table 2 indicates that as the number of bounce increases, smaller angular separation (capable of launching more number of rays) is required for the convergence. In fact, the variation in the number of rays w.r.t. number of bounce does not affect the RF field computation (Choudhury et al. 2013).

The cumulative ray path data is generated for EM field computation till 40 bounces. The number of rays that reach the receiver w.r.t. N-bounce and the program execution time is given in Table 3. It can be seen from Tables 2 and 3 that, for 0.1°, the exact convergence achieved is till four bounces only. However, the small variation in number of rays with increase in angular separation does not effect in RF field computation.

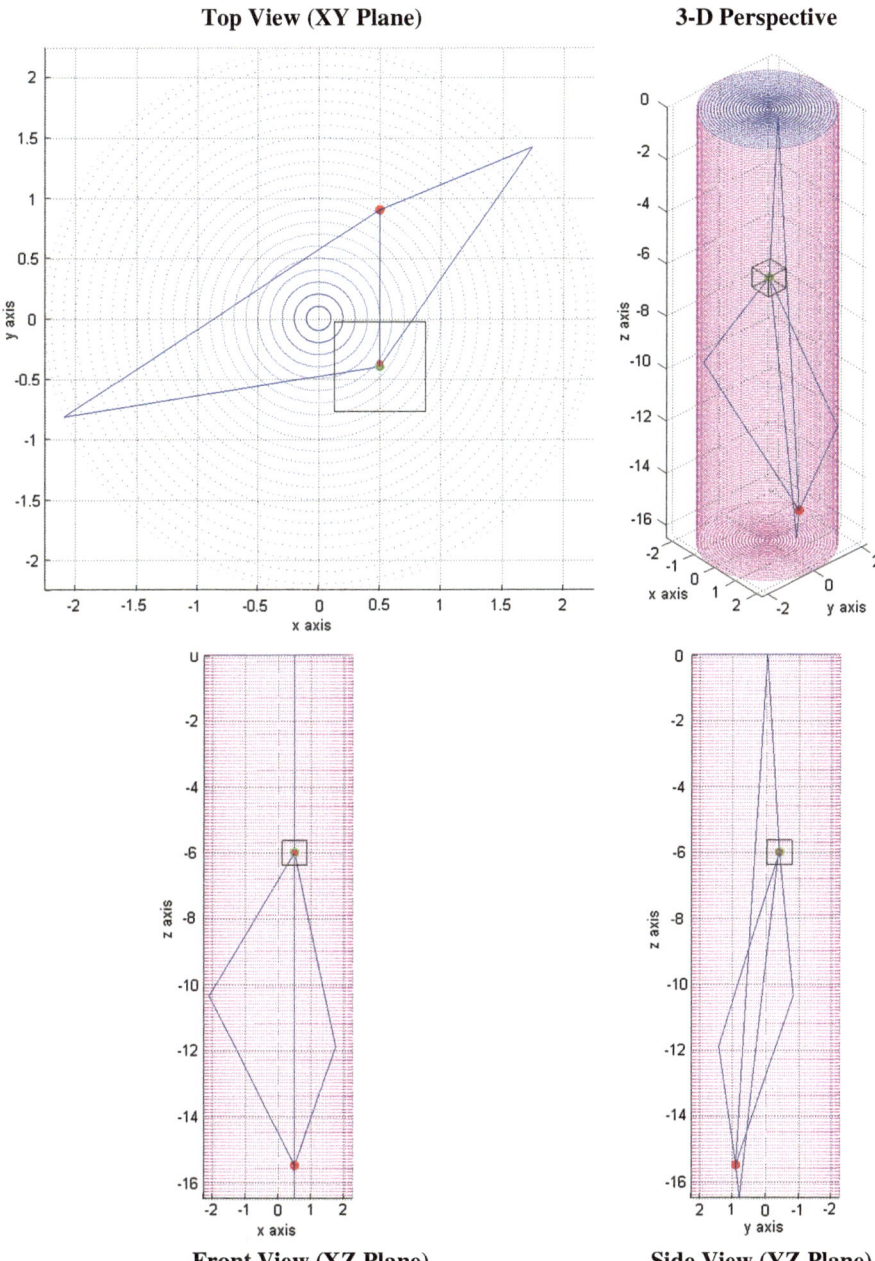

Fig. 17 Ray path of rays that reach the receiver after *single bounce* (*Red dot* represents the source point and *green dot* represents the center of the reception sub-cube)

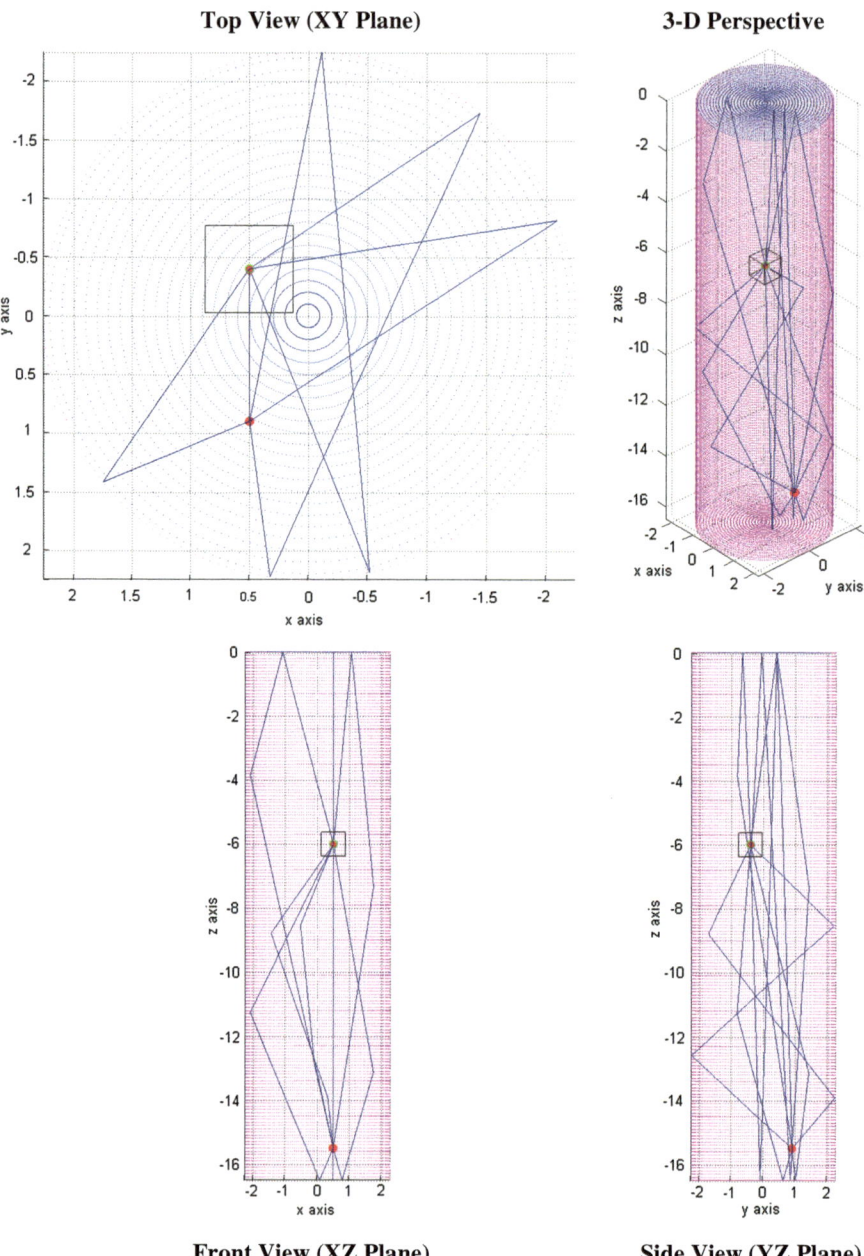

Fig. 18 Ray path of a rays that reach the receiver after *two bounces* (*Red dot* represents the source point and *green dot* represents the center of the reception sub-cube)

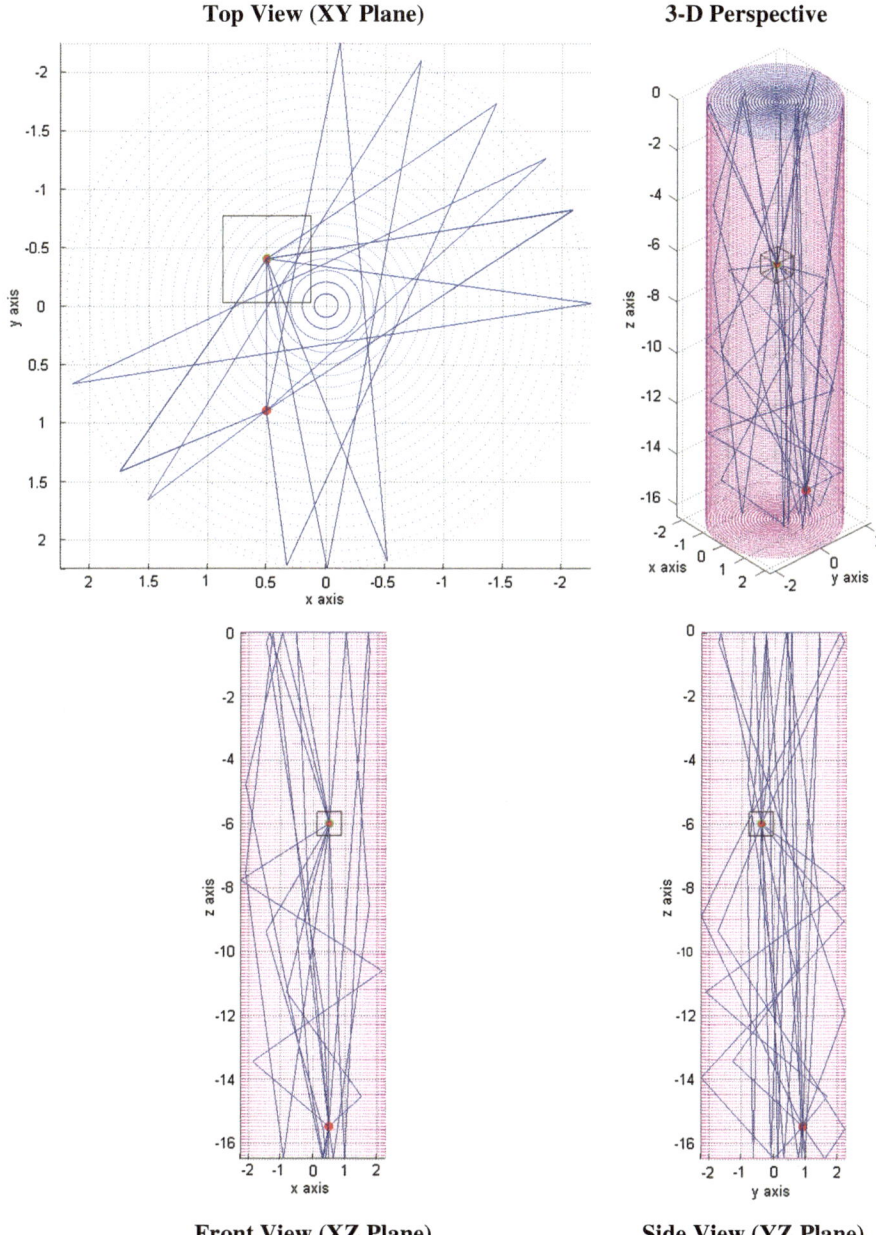

Fig. 19 Ray path of a rays that reach the receiver after *three bounces* (*Red dot* represents the source point and *green dot* represents the center of the reception sub-cube)

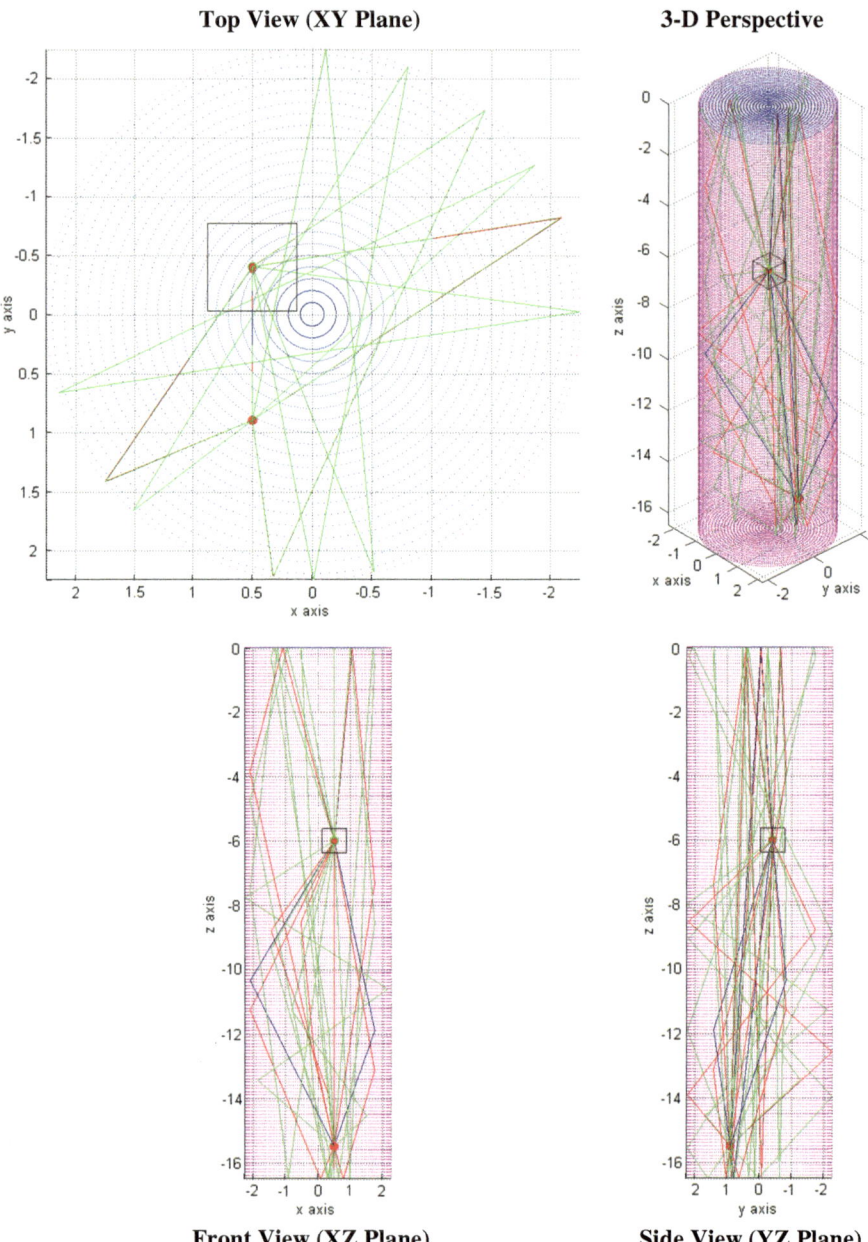

Fig. 20 *Cumulative rays* that reach the receiver up to *three bounces,* visualized in *Matlab* (*Red dot* represents the source point and *green dot* represents the center of the reception sub-cube)

Table 2 Convergence study of rays reaching at N-bounce w.r.t. angular separation

Angular separation	No. of rays launched	1b	2b	3b	4b	5b	6b	7b	8b	9b	10b
1	41,345	4	6	8	5	7	9	12	12	18	24
0.5	165,197	4	8	9	8	11	13	22	28	28	34
0.3	458,675	4	7	10	13	19	22	30	34	41	33
0.2	1,031,769	4	8	11	13	20	24	36	40	51	54
0.1	4,126,183	4	8	12	16	21	32	42	54	58	65
0.05	16,503,013	4	8	12	16	22	32	43	53	64	70
0.03	45,839,676	4	8	12	16	22	31	43	55	66	70

Table 3 Rays reaching receiver cumulatively up to N-bounce and the execution time (4,126,183 rays are launched)

N-bounce	No. of rays received	Program execution time
1b bounce	4	16 s
2b bounce	12	35 s
3b bounce	24	56 s
5b bounce	61	1.53 min
6b bounce	92	2.29 min
8b bounce	188	3.54 min
9b bounce	246	4.42 min
10 bounce	311	5.36 min
15 bounce	844	11.05 min
20 bounce	1745	18.39 min
25 bounce	3025	28.07 min
30 bounce	4631	39.30 min
35 bounce	6576	53.02 min

The minimum and maximum time taken by the rays to reach the receiver w.r.t. N-bounce (individual) is given in Table 4, and Fig. 21 gives the time plot of the same. It is observed that for concave cylindrical structure, the variation in minimum execution time w.r.t. the number of bounce is in significant because of the rays that take helical path to reach the receiver.

4.2   Ray Tracing Inside the GPOR

After modeling the GPOR, a transmitter and a receiver are placed randomly, say at S(0.5, 0.9, −15.5) and R(0.5, −0.4, −6.0) inside the GPOR. As explained in refined ray tracing technique in Sect. 3, the first intersection point is calculated using intersection formula between the line and the surface of GPOR. As the structure has two surfaces at different heights, the z-coordinate of the first intersection point is checked and according to the surface, the equation is adopted for calculation of *first incident point*. The corresponding two surfaces are given below:

Table 4 The minimum and maximum time taken by the individual ray w.r.t. N-bounce

Bounce	T_{min} (µs)	T_{max} (µs)
1b	0.0653466680	0.1268980075
2b	0.0638183534	0.1967586151
3b	0.0715235984	0.2367341480
5b	0.0868415081	0.3467050158
6b	0.0993129524	0.4166967207
8b	0.1207610062	0.4571849421
9b	0.1356777812	0.4638421658
10b	0.1495935253	0.5337601996
15b	0.2172636731	0.7934355419
20b	0.2838526065	1.077291538
25b	0.3524848718	1.406720697
30b	0.4220923555	1.633453484
35b	0.4996643180	1.629832057
40b	0.5626530163	2.176839479

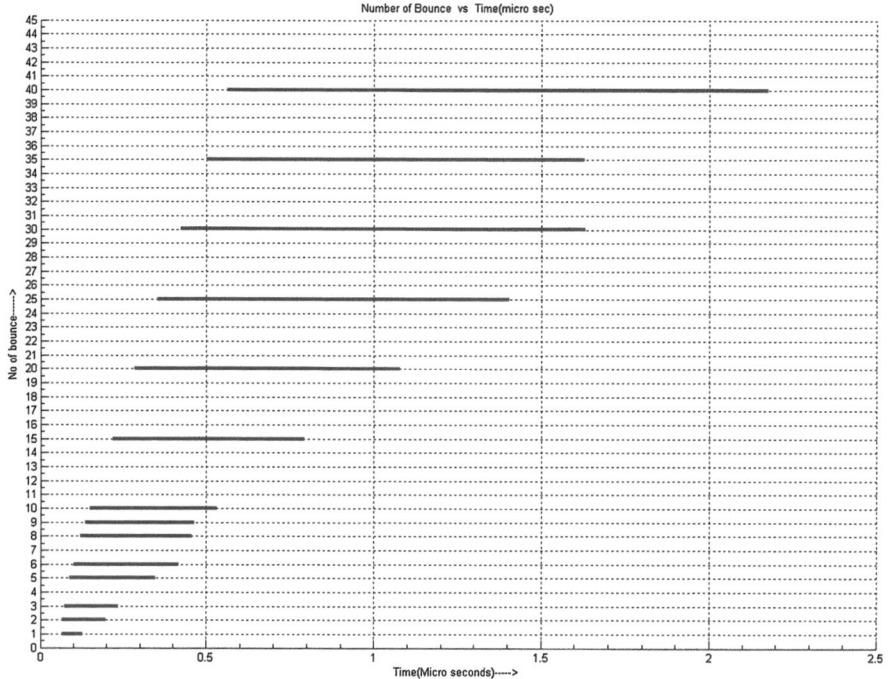

Fig. 21 T_{min}–T_{max} plot w.r.t. number of bounces for right circular cylinder

(a) If $0 \geq z > -16.5$, surface is a GPOR; and
(b) If $z = -16.5$, surface is a plane.

Then the normal at the first incident point is determined by taking the surface normal equation of the corresponding surface (Jha and Wiesbeck 1995). The surface normal of the general paraboloid of revolution (Fig. 22) at a given point is given by (Jha and Wiesbeck 1995)

$$\hat{N} = x_N \hat{i} + y_N \hat{j} + z_N \hat{k} \tag{4}$$

where,

$$x_N = \frac{2u \cos \phi}{\sqrt{a^2 + 4u^2}} \tag{4.a}$$

$$y_N = \frac{2u \sin \phi}{\sqrt{a^2 + 4u^2}} \tag{4.b}$$

$$z_N = \frac{a}{\sqrt{a^2 + 4u^2}} \tag{4.c}$$

where a and u are the shaping parameters of the GPOR.

The ray path details are calculated using the refined ray tracing algorithm.

4.2.1 Results and Visualization of Ray Path

The ray path propagation inside the GPOR is visualized using *Matlab*. Table 5 represents only 42 rays that reach the receiver cumulatively up to three bounces (excluding the direct ray) out of 4,126,183 launched rays. There are two rays, which reach after one bounce, 10 rays after two bounces, and 30 rays after three bounces. Figures 23, 24, and 25 give the visualization of one-bounce, two-bounce, and three-bounce rays reaching receiver. Figure 26 gives the cumulative number of rays till the third bounce.

4.2.2 Discussion

The minimum and maximum time taken by the rays to reach the receiver w.r.t. N-bounce (individual) is given in Table 6, and Fig. 27 gives the time plot of the same.

4.3 Ray Tracing Inside GPOR Frustum

After analytical surface modeling of GPOR frustum (as described in Sect. 2 of this book), a transmitter and a receiver are placed randomly at S(0.5, 0.9, −15.5) and R

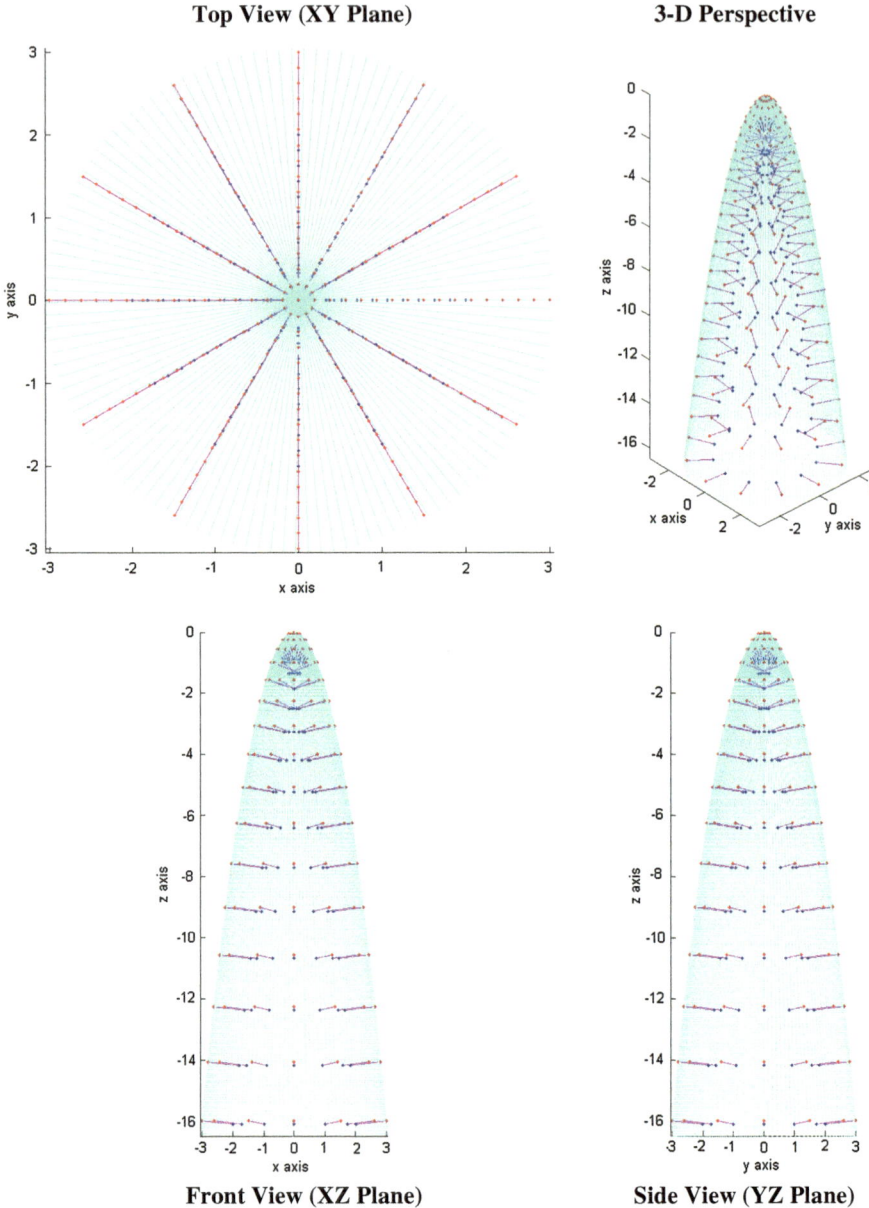

Fig. 22 Surface normal visualization of GPOR at different incident points

(0.5, −0.4, −6.0). As explained earlier, the first intersection point is calculated using intersection formula between the line and the surface of GPOR. As the structure has three surfaces at different heights, the z-coordinate of the first intersection point is

Table 5 Description of 42 rays that reached the receiver (till 3 bounce) out of 4,126,183 rays launched at 0.1° angular separation

Sl. No	θ	φ	Ex	Ey	Ez	Path length (m)	Time (µs)
1	41.452	272.061	1.473	−0.216	−1.602	17.087	0.056
2	34.700	324.465	1.445	−0.298	−1.594	17.113	0.057
3	45.523	255.027	−2.562	−0.261	−2.341	17.959	0.059
4	39.720	207.822	−2.504	−0.388	−2.248	18.004	0.060
5	36.593	205.495	−1.048	−1.754	−2.787	18.261	0.060
6	42.609	204.082	1.389	−0.117	−0.820	18.305	0.061
7	43.153	274.344	0.472	−1.568	−2.438	18.414	0.061
8	46.715	253.185	−1.164	−1.674	−2.435	18.502	0.061
9	32.265	325.997	0.341	−1.693	−3.051	18.548	0.061
10	43.427	158.151	1.384	−0.162	−0.872	18.837	0.062
...							
...							
...							
35	18.287	248.937	1.849	0.025	−4.066	38.833	0.129
36	34.103	263.697	−0.546	−0.147	5.462	38.921	0.129
37	27.923	262.835	−0.028	0.488	5.590	39.038	0.130
38	176.601	240.217	−0.517	0.399	5.999	40.162	0.133
39	24.242	147.734	−0.506	0.456	5.994	40.182	0.133
40	14.738	262.553	−0.167	−0.650	−10.499	48.010	0.160
41	166.549	262.598	−0.179	−0.719	−10.499	48.801	0.1626
42	13.327	143.555	−0.343	−0.278	−10.500	54.622	0.182

checked and according to the surface, the equation is adopted for calculation of *first incident point*. The corresponding three surfaces are given below:

(a) At $z = -4$, surface is a plane;
(b) If $-4 \geq z > -16.5$, surface is a GPOR; and
(c) If $z = -16.5$, surface is a plane.

Then the normal at the first incident point is by determined taking the normal equation of the GPOR surface (Jha and Wiesbeck 1995). The surface normal of the general paraboloid of revolution (Fig. 28) at a given point is given as

$$\hat{N} = x_N\hat{i} + y_N\hat{j} + z_N\hat{k} \tag{5}$$

where,

$$x_n = \frac{2u \cos \phi}{\sqrt{a^2 + 4u^2}} \tag{5.a}$$

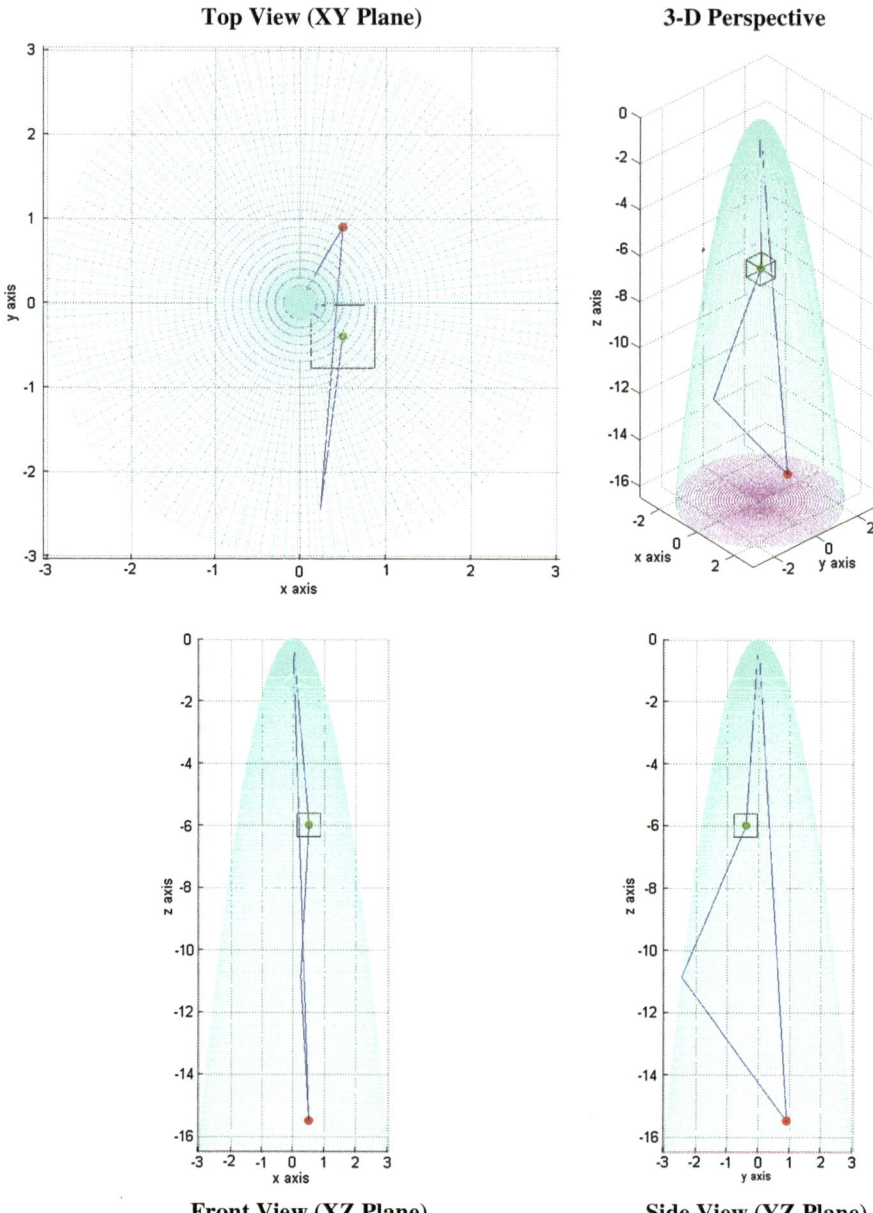

Fig. 23 Ray path of a ray that reach the receiver after *single bounce* (*Red dot* represents the source point and *green dot* represents the center of the reception sub-cube)

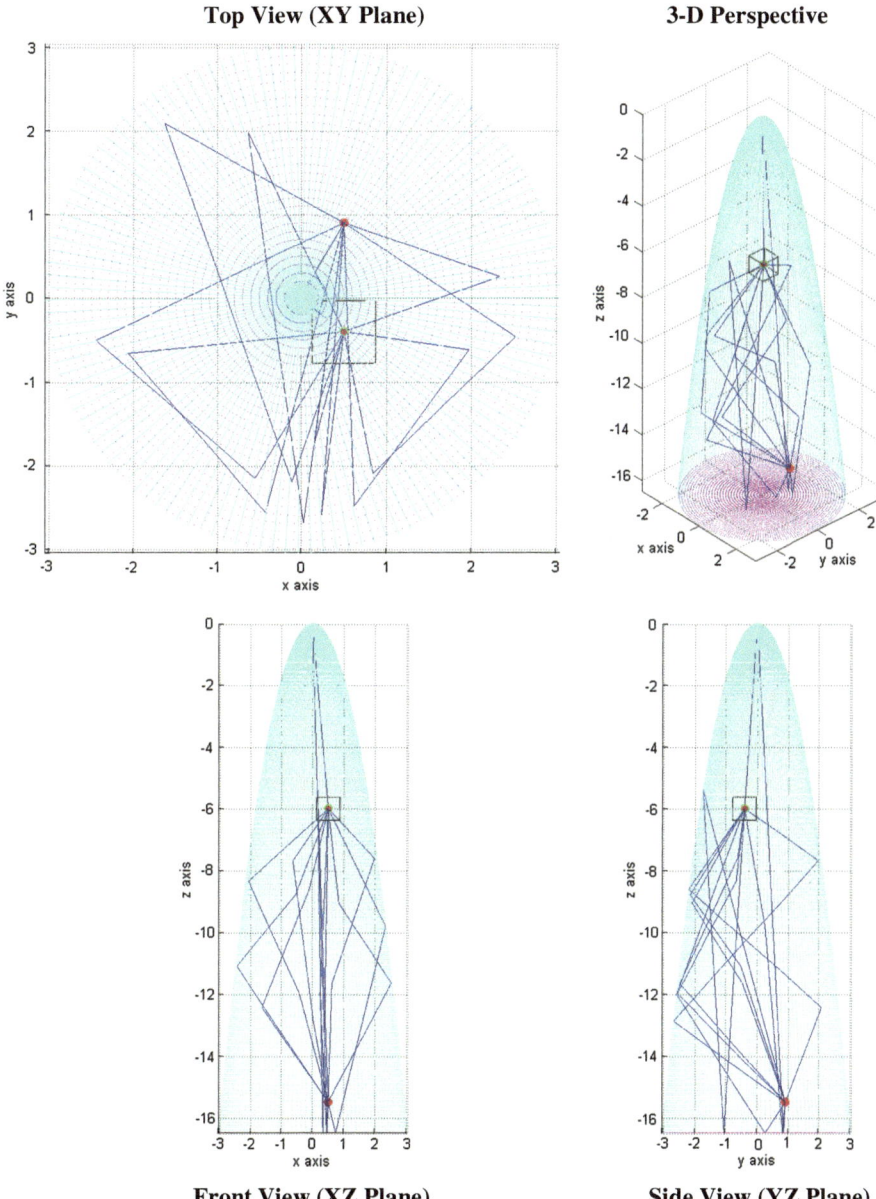

Fig. 24 Ray path of rays that reach the receiver after *two bounces* (*Red dot* represents the source point and *green dot* represents the center of the reception sub-cube)

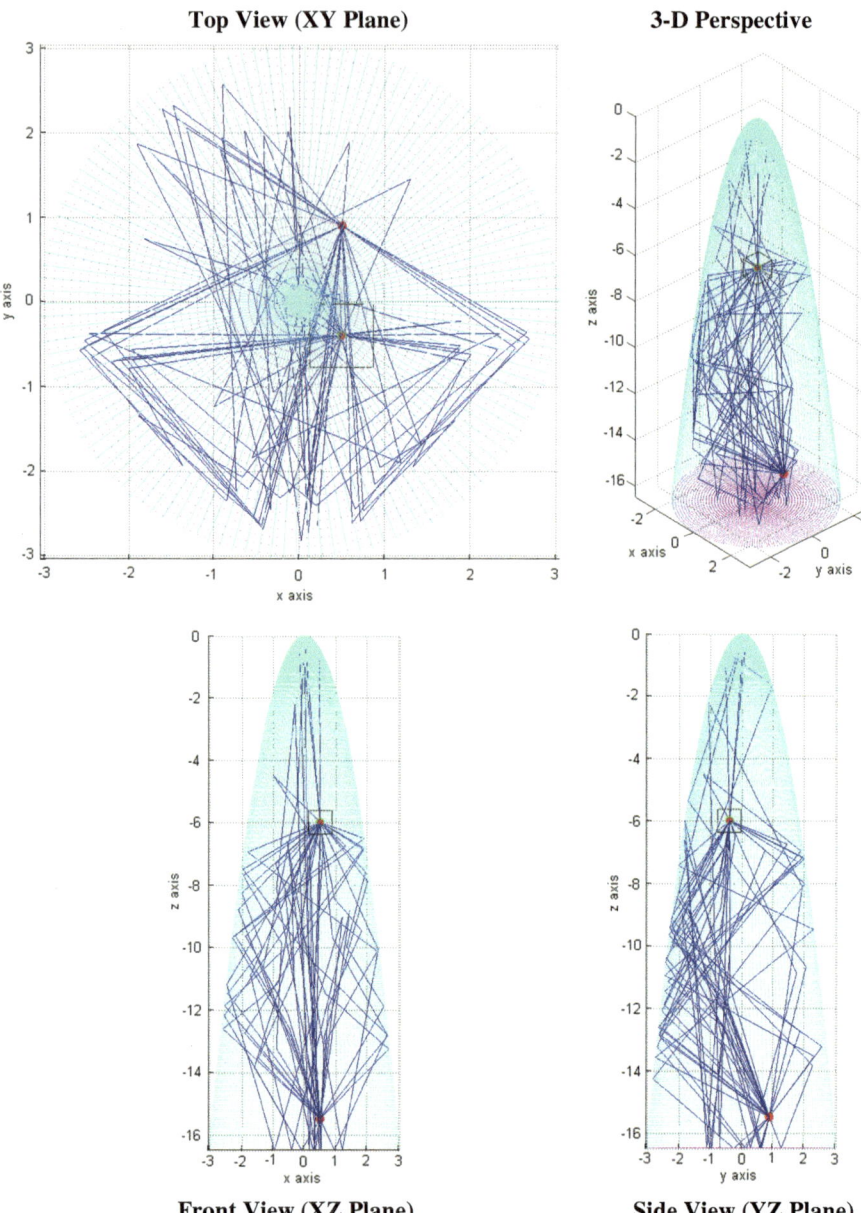

Fig. 25 Ray path of rays that reach the receiver after *three bounces* (*Red dot* represents the source point and *green dot* represents the center of the reception sub-cube)

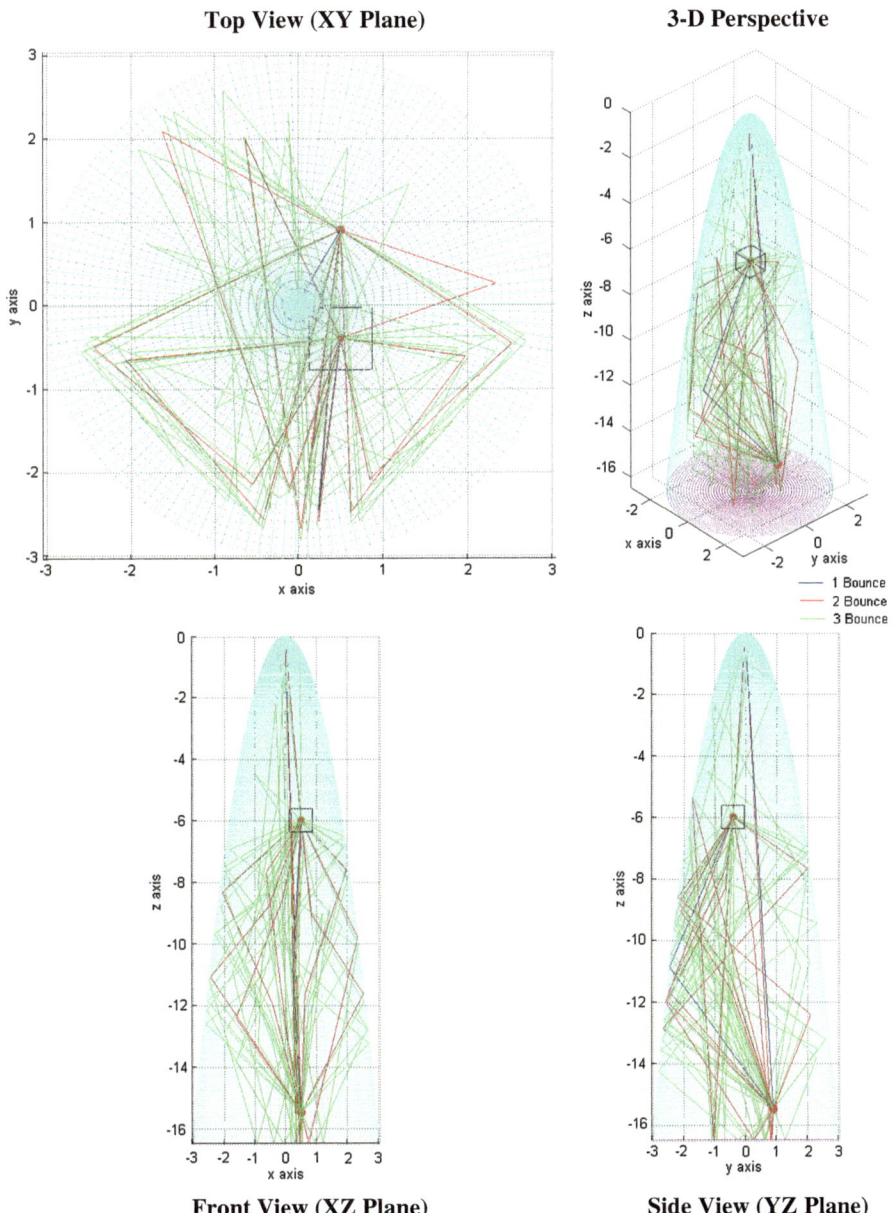

Fig. 26 *Cumulative rays* that reach the receiver up to *three bounces* visualized in *Matlab* (*Red dot* represents the source point and *green dot* represents the center of the reception sub-cube)

Table 6 The minimum and maximum time taken by the individual ray w.r.t. N-bounce

Bounces	$T_{\min}$ (μs)	$T_{\max}$ (μs)
1b	0.06653254	0.12722019
2b	0.05695799	0.16003386
3b	0.05704445	0.18207432
4b	0.05774340	0.19818648
5b	0.06353398	0.19001644
6b	0.06647436	0.21749209
7b	0.07210666	0.28821882
8b	0.08266083	0.21083427
9b	0.09052185	0.19532442
10b	0.09768129	0.19112064
11b	0.10758060	0.19399442
12b	0.12331454	0.21015567
13b	0.13465425	0.16686257
14b	0.14841293	0.17886249
15b	0.13307072	0.14980707

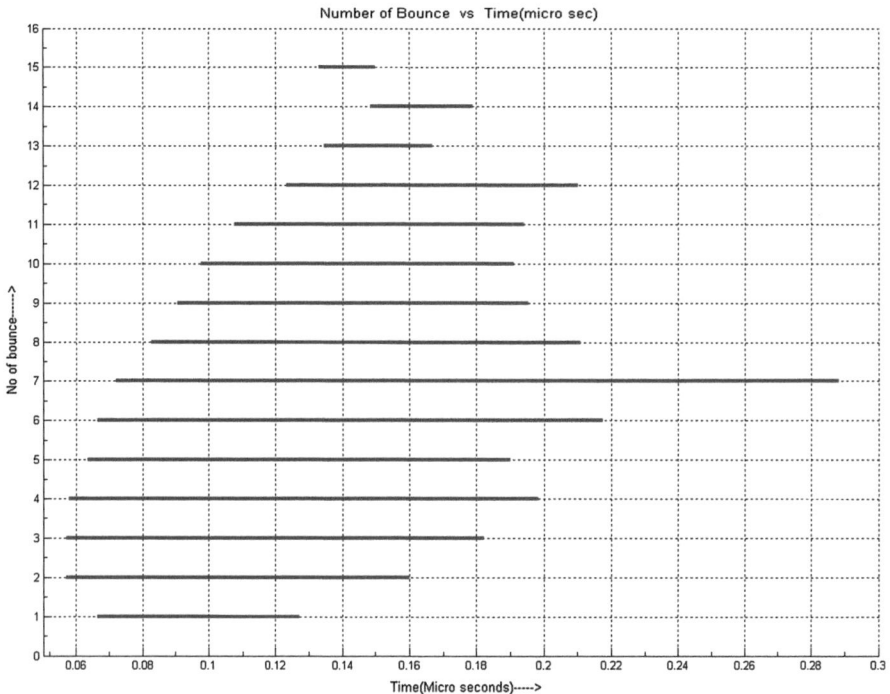

Fig. 27 $T_{\min}$–$T_{\max}$ plot w.r.t. number of bounces for GPOR

Top View (XY Plane)   **3-D Perspective**

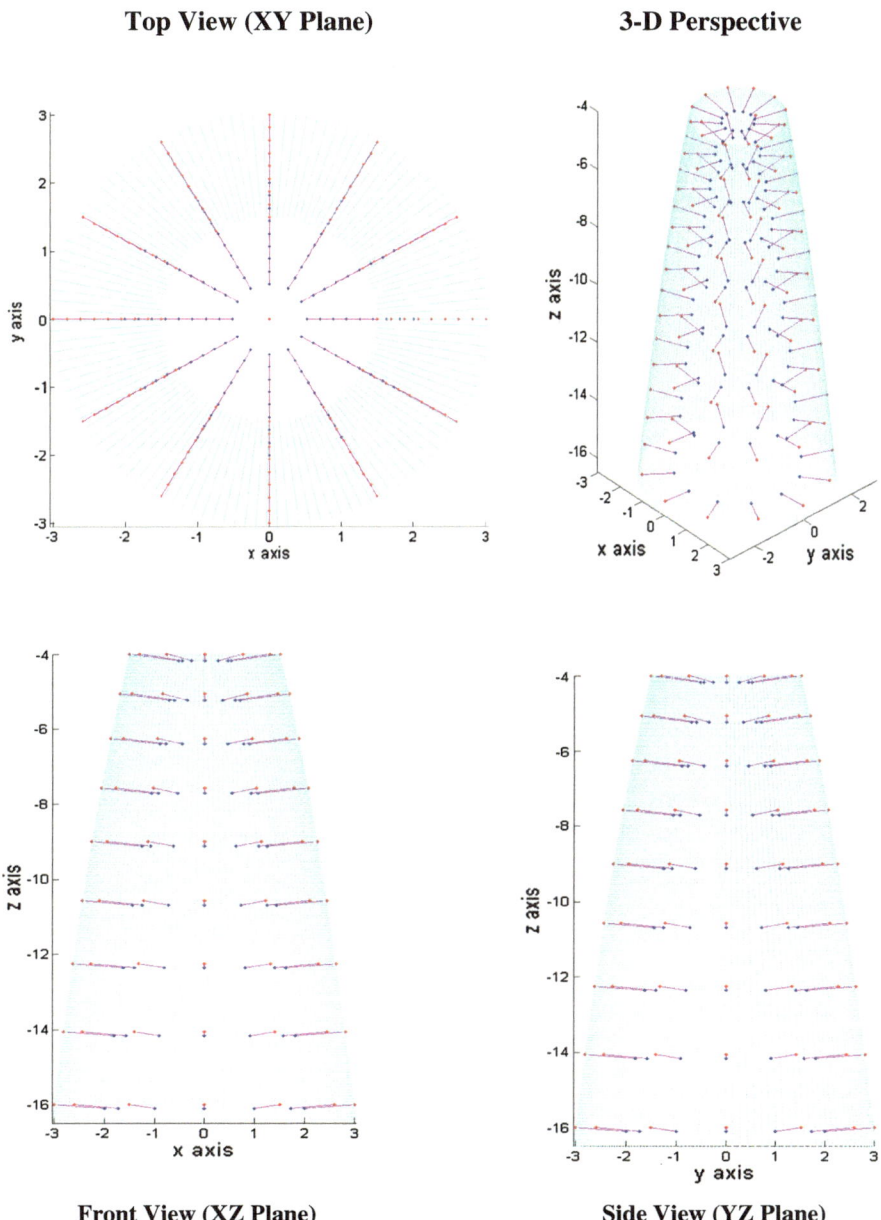

Front View (XZ Plane)   **Side View (YZ Plane)**

Fig. 28 Surface normal visualization of GPOR Frustum at different incident points

$$y_n = \frac{2u \sin \phi}{\sqrt{a^2 + 4u^2}} \tag{5.b}$$

$$z_n = \frac{a}{\sqrt{a^2 + 4u^2}} \tag{5.c}$$

where a and u are the shaping parameters of the GPOR.

The ray path details are calculated using the refined ray tracing algorithm as explained in Sect. 3.

4.3.1 Results and Visualization of Ray Path

The ray path propagation inside the manned GPOR frustum is visualized using *Matlab*. Table 7 represents the 46 rays that reach the receiver cumulatively up to three bounces (excluding the direct ray) out of 4,126,183 launched rays. There are four rays, which reach after one bounce, 14 rays after two bounces, and 28 rays

Table 7 Description of 46 rays that reached the receiver (till 3 bounce) out of 4,126,183 rays launched at 0.1° angular separation

Sl. no	θ	ϕ	Ex	Ey	Ez	Path length (m)	Time (µs)
1	16.600	345.13	1.757	0.831	−3.399	16.501	0.055
2	41.399	272.034	1.456	−0.222	−1.589	16.857	0.056
3	32.249	325.994	0.341	−1.691	−3.049	17.358	0.057
4	45.499	255.029	−2.558	−0.260	−2.339	17.958	0.059
5	36.594	205.495	−1.048	−1.754	−2.787	18.261	0.060
6	42.563	204.086	1.382	−0.117	−0.817	18.301	0.061
7	36.071	265.561	−0.260	−2.061	−4.871	18.492	0.061
8	165.800	340.419	1.837	0.650	−3.800	18.732	0.062
9	38.492	150.857	−0.612	−1.798	−2.606	18.780	0.062
10	142.673	271.983	1.512	−0.220	−1.886	18.795	0.062
...							
...							
...							
36	13.332	187.606	1.217	−1.070	−2.971	26.871	0.089
37	28.198	265.268	−0.070	−0.506	2.000	27.449	0.091
38	20.533	322.736	0.159	−0.585	1.999	27.698	0.092
39	175.206	269.994	1.62E−05	0.167	2.000	28.098	0.093
40	153.645	265.472	−0.060	−0.468	2.000	29.370	0.097
41	14.753	262.547	−0.167	−0.645	−10.500	44.607	0.148
42	166.537	262.637	−0.176	−0.715	−10.499	45.412	0.151
43	2.151	269.939	0.000	0.394	−10.500	47.033	0.156
44	8.599	194.239	−3.556	0.408	−10.374	48.464	0.161
45	178.005	270.058	−0.000	0.365	−10.501	49.030	0.163
46	1.933	270.000	−3.3E−07	0.067	2	51.029	0.170

after three bounces. Figures 29, 30, and 31 give the visualization of one-bounce, two-bounce, and three-bounce rays reaching receiver. Figure 32 gives the cumulative number of rays till the third bounce.

4.3.2 Discussion

The cumulative ray path data is generated for EM field computation till 20 bounces. The number of rays that reach the receiver w.r.t. N-bounce and the program execution time is given in Table 8. In the case of GPOR frustum because of the curvature, there are rays, which traverse through the surface. Hence, the rays are dropped here as the reflected ray path is considered. The minimum and maximum time taken by the rays to reach the receiver w.r.t. N-bounce is given in Table 9, and Fig. 33 gives the time plot of the same.

4.4 Ray Tracing Inside the GPOR and Right Circular Cylinder

After modeling the hybrid structure of GPOR and right circular cylinder, a transmitter and a receiver are placed randomly at S(0.5, 0.9, −15.5) and R(0.5, −0.4, −6.0) inside the GPOR–right circular cylinder. As explained in refined ray tracing technique in Sect. 3, the first intersection point is calculated using intersection formula between the line and the corresponding surface. As the hybrid structure has three surfaces at different heights, the z-coordinate of the first intersection point is checked and according to the surface, the equation is adopted for calculation of *first incident point*. The corresponding four surfaces are given below:
At $z = 0$, surface is a plane

(a) If $0 \geq z > -9$, surface is a gpor;
(b) If $-9 \geq z > -16.5$, surface is a right circular cylinder; and
(c) If $z = -16.5$, surface is a plane.

Then the normal at the first incident point is determined by taking the normal equation of the corresponding surfaces (Jha and Wiesbeck 1995)

$$\hat{N} = x_N \hat{i} + y_N \hat{j} + z_N \hat{k} \tag{6}$$

The normal at the right circular cylinder is given by Jha and Wiesbeck (1995)

$$x_n = a \cos \phi \tag{6.a}$$

$$y_n = a \sin \phi \tag{6.b}$$

$$z_n = 0 \tag{6.c}$$

where a is the radius of the right circular cylinder.

Top View (XY Plane)   **3-D Perspective**

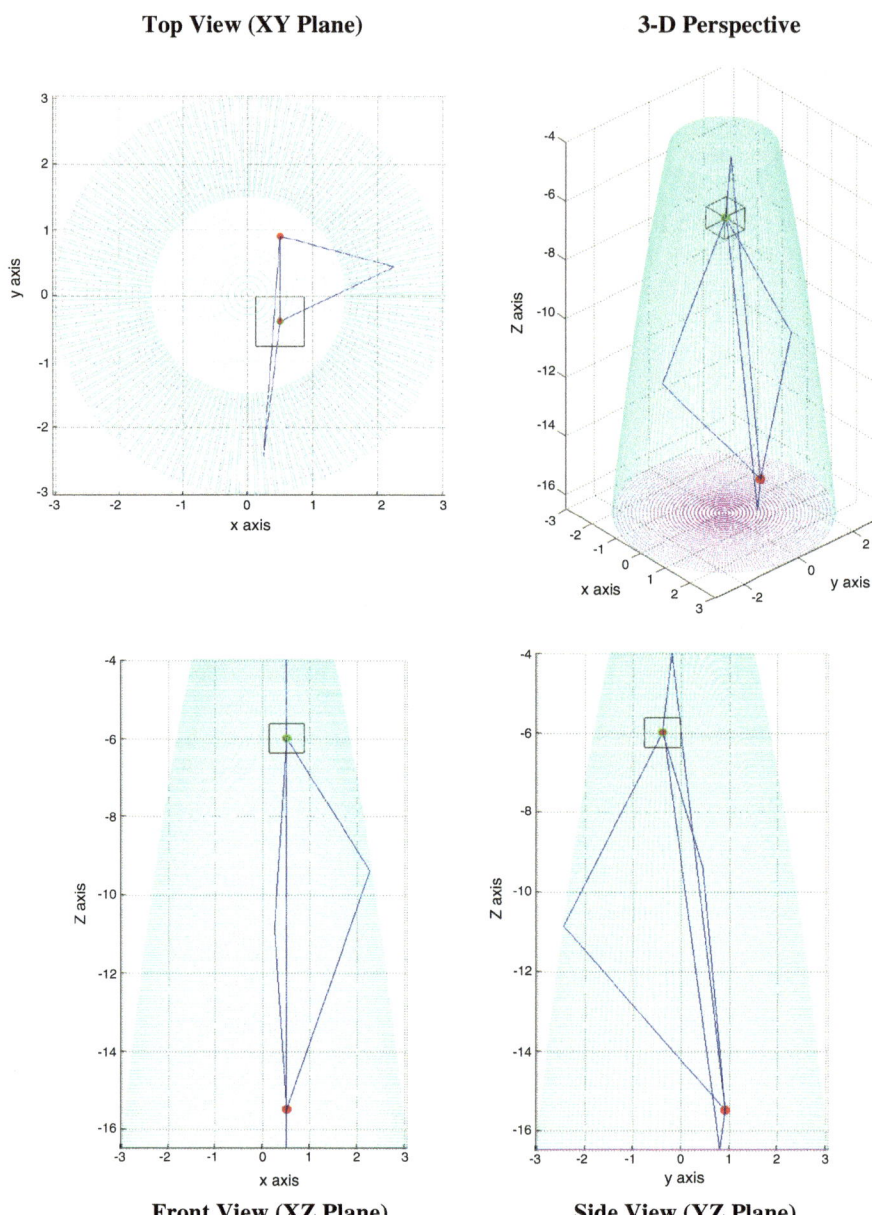

Front View (XZ Plane)   **Side View (YZ Plane)**

Fig. 29 Ray path of rays that reach the receiver after *single bounce* (*Red dot* represents the source point and *green dot* represents the center of the reception sub-cube)

Top View (XY Plane)   **3-D Perspective**

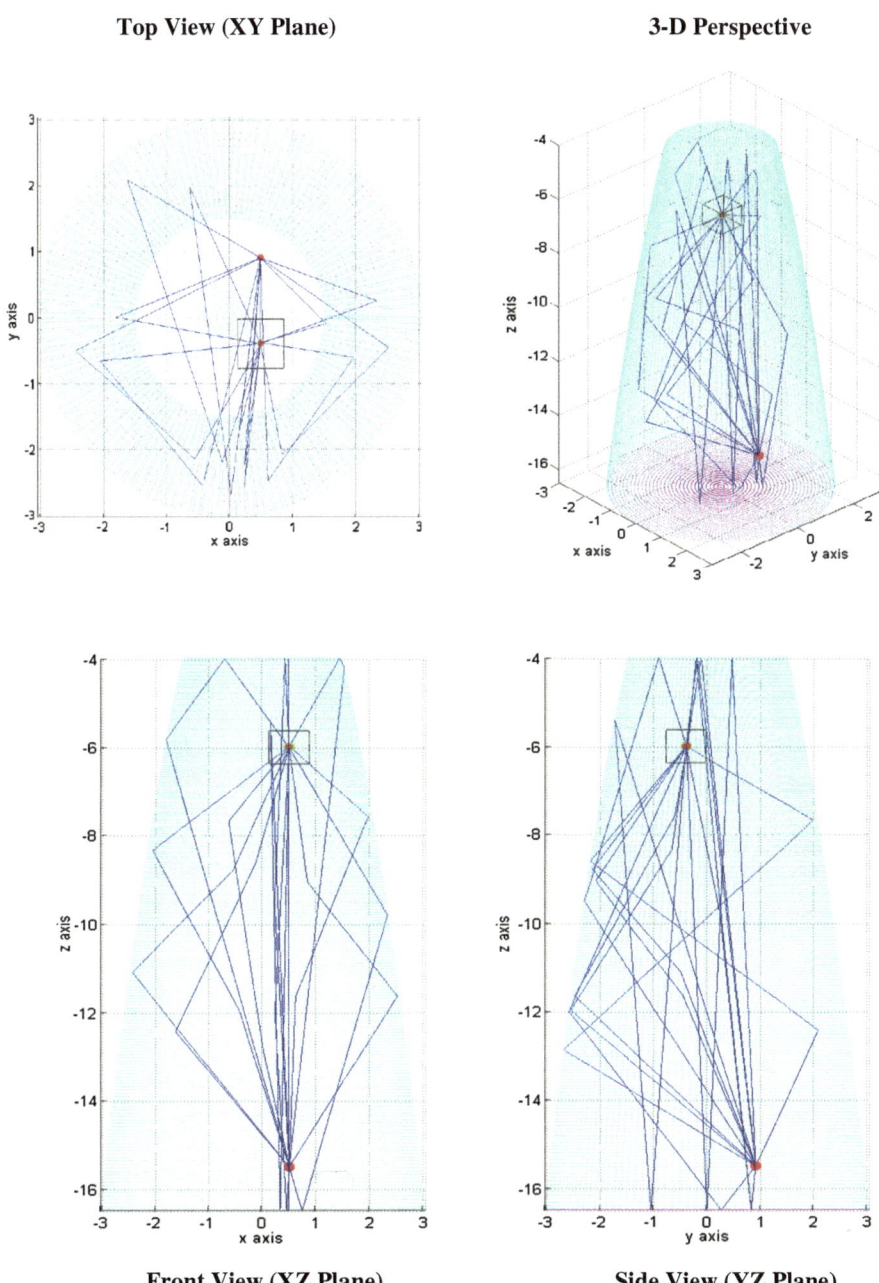

Front View (XZ Plane)   **Side View (YZ Plane)**

Fig. 30 Ray path of rays that reach the receiver after *two bounces* (*Red dot* represents the source point and *green dot* represents the center of the reception sub-cube)

Top View (XY Plane)   **3-D Perspective**

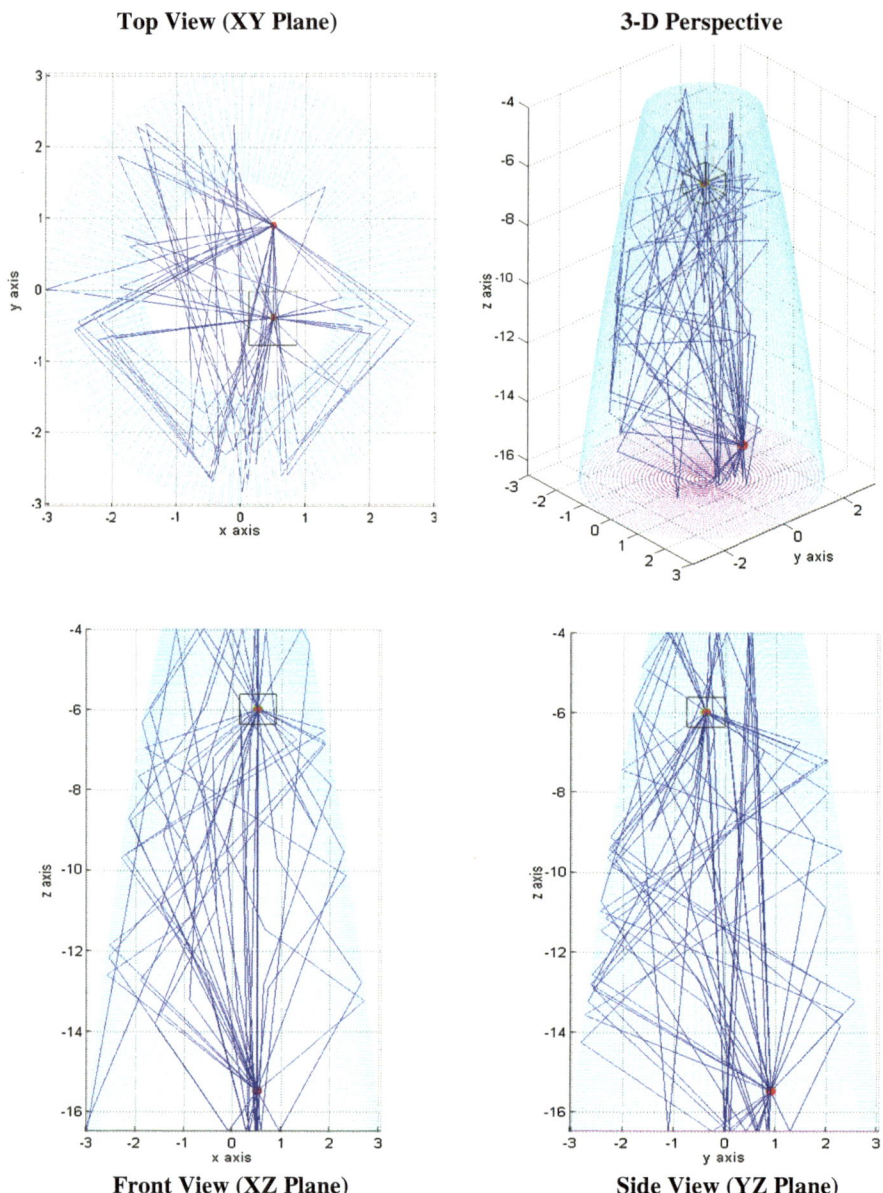

Front View (XZ Plane)   **Side View (YZ Plane)**

Fig. 31 Ray path of rays that reach the receiver after *three bounces* (*Red dot* represents the source point and *green dot* represents the center of the reception sub-cube)

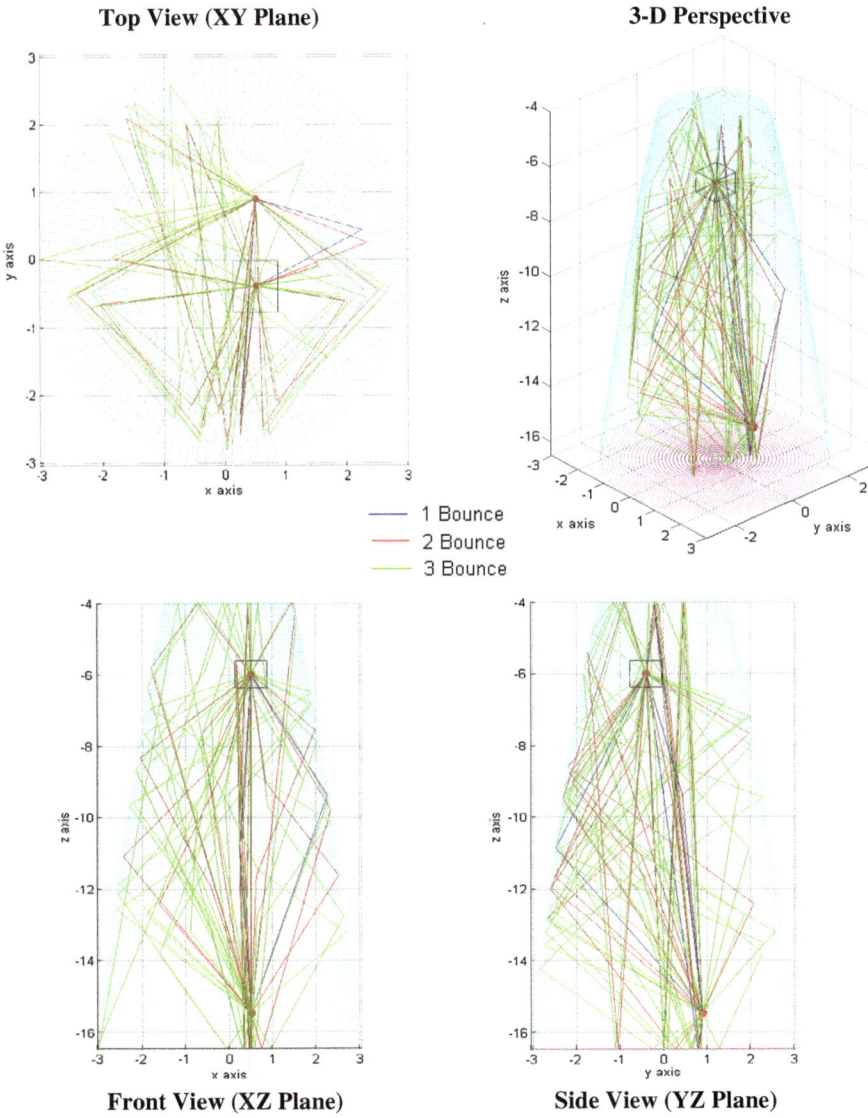

Top View (XY Plane)   **3-D Perspective**

— 1 Bounce
— 2 Bounce
— 3 Bounce

Front View (XZ Plane)   **Side View (YZ Plane)**

Fig. 32 *Cumulative rays* that reach the receiver up to *three bounces* visualized in *Matlab* (*Red dot* represents the source point and *green dot* represents the center of the reception sub-cube)

The surface normal of the general paraboloid of revolution at a given point is given by

$$x_n = \frac{2u \cos \phi}{\sqrt{a^2 + 4u^2}}$$ (6.d)

Table 8 Rays reaching the receiver cumulatively up to N-bounce and the execution time (4,126,183 rays are launched)

N-bounce	No. of rays received	Program ex ecution time
1 bounce	4	22 s
2 bounce	18	42 s
3 bounce	46	1.06 min
5 bounce	148	2.02 min
6 bounce	198	2.36 min
8 bounce	286	3.38 min
9 bounce	309	4.08 min
10 bounce	326	4.31 min
15 bounce	355	6.54 min
20 bounce	359	9.13 min

Table 9 The minimum and maximum time taken by the rays to reach the receiver w.r. t. N-bounce

Bounce	T_{min} (µs)	T_{max} (µs)
1b	0.055006	0.087067
2b	0.056192	0.156777
3b	0.061006	0.170096
4b	0.065698	0.176756
5b	0.064762	0.253386
6b	0.069621	0.254249
7b	0.080447	0.336703
8b	0.096905	0.254938
9b	0.099582	0.328168
10b	0.108911	0.300373
15b	0.132559	0.349070

$$y_n = \frac{2u \sin \phi}{\sqrt{a^2 + 4u^2}} \tag{6.e}$$

$$z_n = \frac{a}{\sqrt{a^2 + 4u^2}} \tag{6.f}$$

where a and u are the shaping parameters of the GPOR. The ray path details are calculated using the refined ray tracing algorithm.

Normal to the hybrid structure of GPOR–Right circular cylinder (Fig. 34) is visualized by the surface normal equation of the right circular cylinder and GPOR described above.

4.4.1 Results and Visualization of Ray Path

The ray path propagation inside the GPOR–right circular cylinder is visualized using *Matlab*. Table 10 shows the 27 rays that reach the receiver cumulatively up to

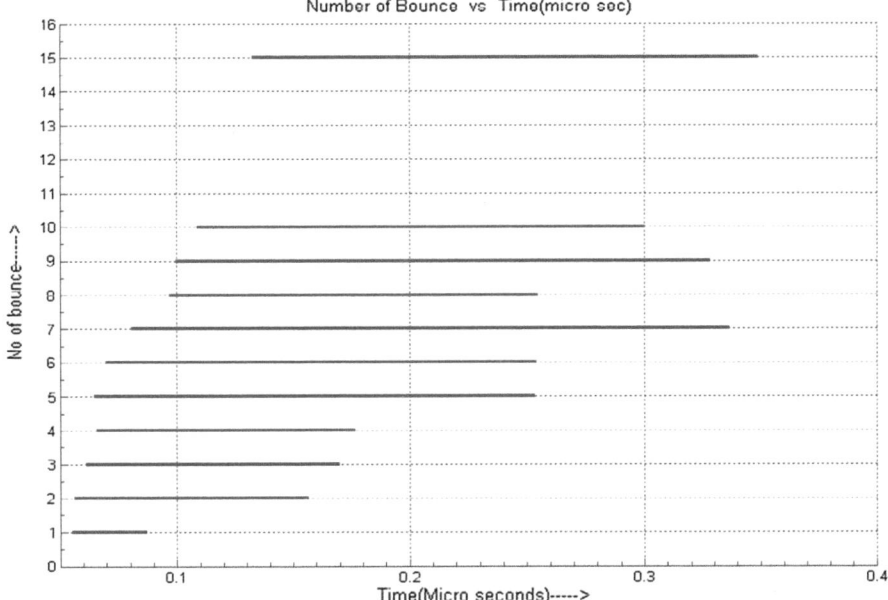

Fig. 33 T_{min}–T_{max} plot w.r.t. number of bounces for GPOR frustum

three bounces excluding the direct ray. There are three rays, which reach after one bounce, six rays after two bounces, and 18 rays after three bounces. Figures 35, 36, and 37 give the visualization of one-bounce, two-bounce, and three-bounce rays reaching receiver. Figure 38 gives the cumulative number of rays till 3rd bounce.

4.4.2   Discussion

The number of rays that reach the receiver w.r.t. N-bounce (cumulatively) and the program execution time of the corresponding bounce is given in Table 11. The minimum and maximum time taken by the rays to reach the receiver w.r.t. N-bounce is given in Table 12, and Fig. 39 gives the time plot of the same.

4.5   Ray Propagation Inside the Space Module

After modeling the Space module, a transmitter and a receiver are placed randomly at S (0.5, 0.9, −15.5) and R (0.5, −0.4, −6.0) inside the space module. As the hybrid structure has four surfaces at different heights, the z-coordinate of the first intersection point is checked and according to the surface, the equation is adopted for calculation of *first incident point*. The corresponding four surfaces are given below:

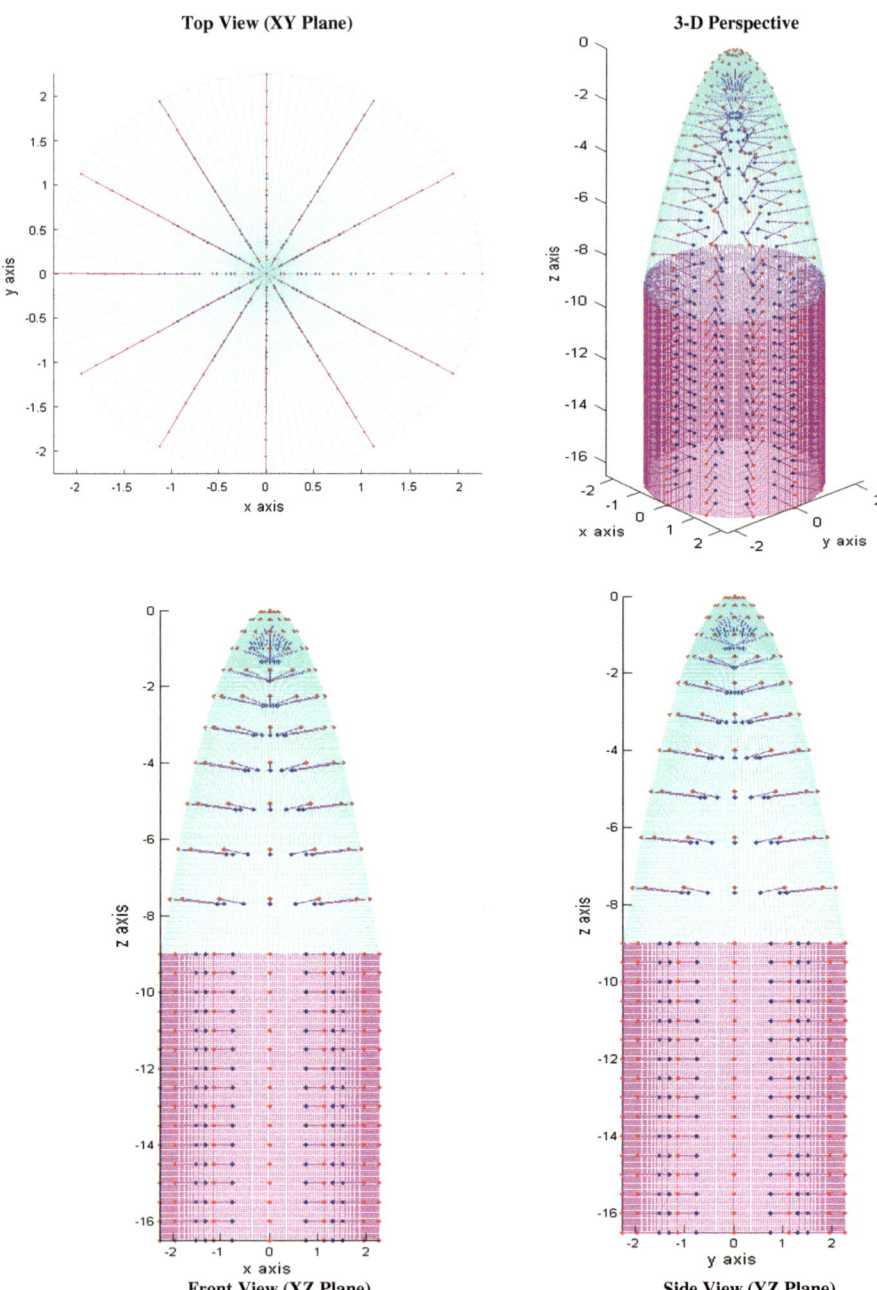

Fig. 34 Surface normal of hybrid structure (GPOR and right circular cylinder) at different incident points

Table 10 Description of 27 rays that reached the receiver (till 3 bounce) out of 4,126,183 rays launched at 0.1° angular separation

Sl. no	Θ	φ	Ex	Ey	Ez	Path length (m)	Time (µs)
1	40.271	212.571	1.467	0.1638	−0.979	17.125	0.057
2	40.451	298.404	−2.566	0.097	−1.753	17.797	0.059
3	31.151	213.503	−2.595	−0.418	−4.349	18.204	0.060
4	43.335	260.575	−1.135	2.367	−1.599	18.333	0.061
5	45.017	307.361	−1.016	−1.625	−1.769	18.868	0.062
6	144.801	212.519	1.501	0.164	−1.219	18.986	0.063
7	46.056	205.749	0.269	−1.501	−1.477	19.048	0.063
8	45.065	298.359	1.380	0.222	−0.342	19.348	0.064
9	20.515	22.613	1.246	1.819	−5.892	19.571	0.065
10	46.273	279.341	1.388	0.387	−0.343	19.615	0.065
...							
...							
...							
20	47.679	214.441	−2.357	0.858	−0.504	21.090	0.070
21	53.994	221.681	−2.469	0.502	−0.916	21.216	0.070
22	137.755	258.964	−1.020	2.588	−3.064	21.253	0.070
23	57.754	251.189	−0.322	−1.697	−1.879	21.924	0.073
24	162.823	22.6133	1.246	1.819	−7.133	22.932	0.076
25	3.834	240.156	−0.523	0.395	5.999	38.167	0.127
26	176.601	240.221	−0.510	0.392	6.000	40.159	0.133
27	3.851	240.104	1.138	−0.985	−10.500	59.399	0.197

(a) At $z = -4$, surface is a plane;
(b) If $-4 > z > -9$, surface is a GPOR;
(c) If $-9 \geq z > -16.5$, surface is a right circular cylinder; and
(d) If $z = -16.5$, surface is a plane.

Then the normal at the first incident point is determined by taking the surface normal equation of the corresponding surfaces (Jha and Wiesbeck 1995)

$$\hat{N} = x_N \hat{i} + y_N \hat{j} + z_N \hat{k} \tag{7}$$

The surface normal at the right circular cylinder is given by Jha and Wiesbeck (1995)

$$x_n = a \cos \phi \tag{7.a}$$

$$y_n = a \sin \phi \tag{7.b}$$

$$z_n = 0 \tag{7.c}$$

where a is the radius of the right circular cylinder.

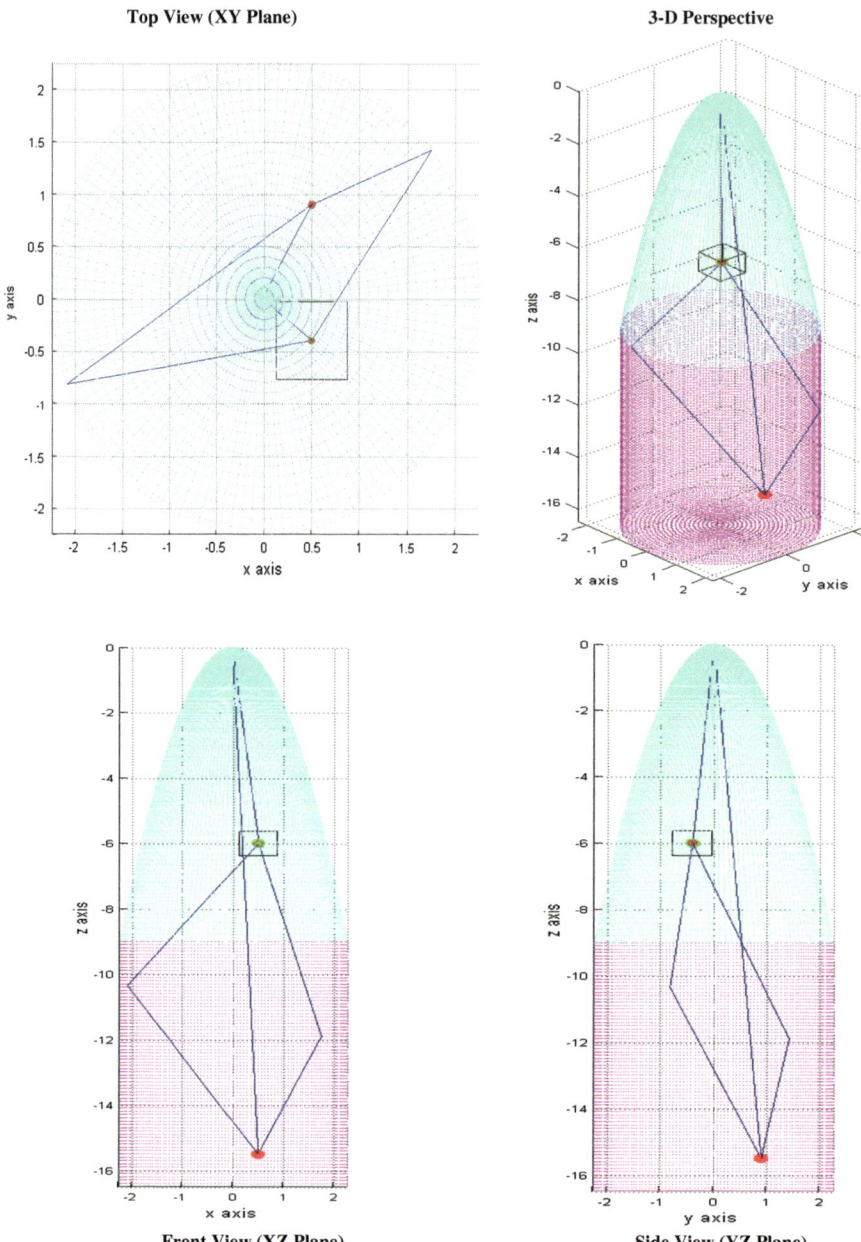

Fig. 35 Ray path of rays that reach the receiver after *single bounce* (*Red dot* represents the source point and *green dot* represents the center of the reception sub-cube)

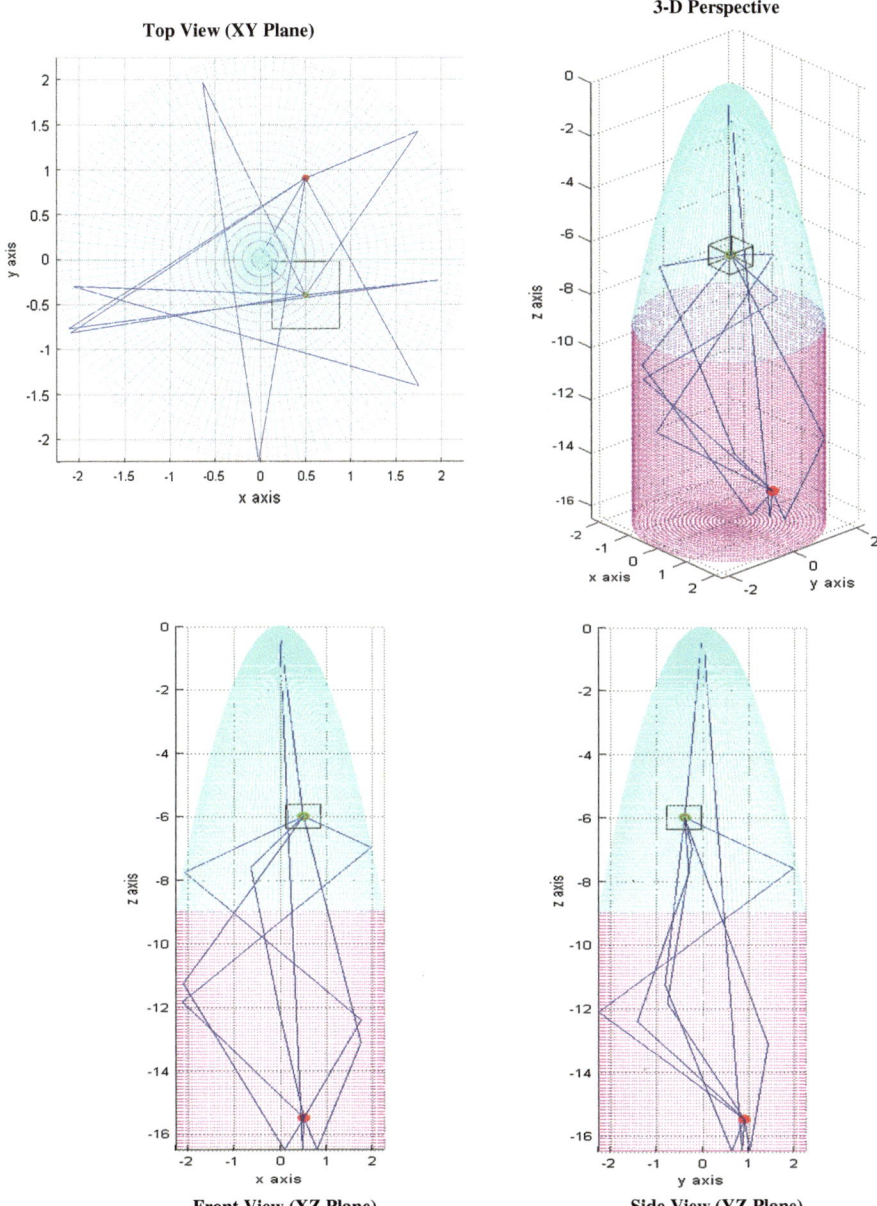

Fig. 36 Ray path of rays that reach the receiver after *two bounces* (*Red dot* represents the source point and *green dot* represents the center of the reception sub-cube)

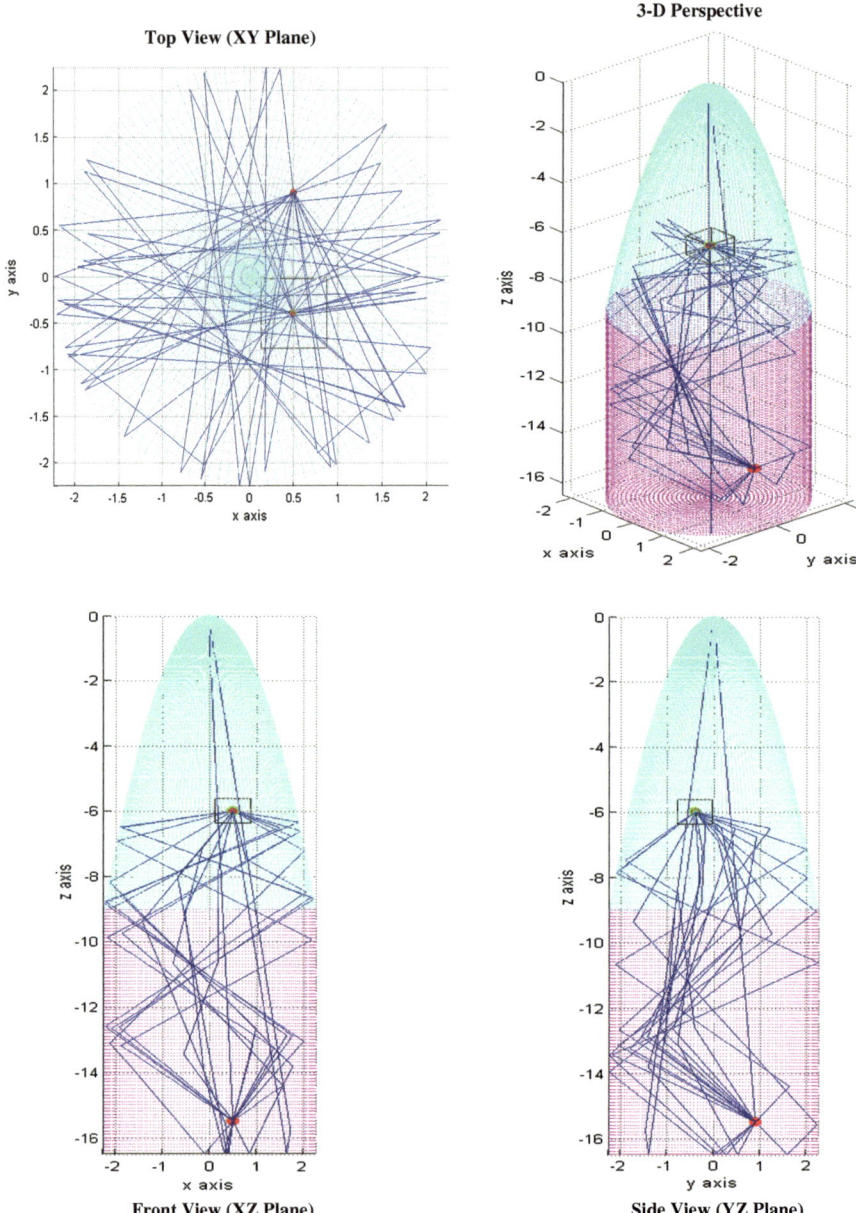

Fig. 37 Ray path of rays that reach the receiver after *three bounces* (*Red dot* represents the source point and *green dot* represents the center of the reception sub-cube)

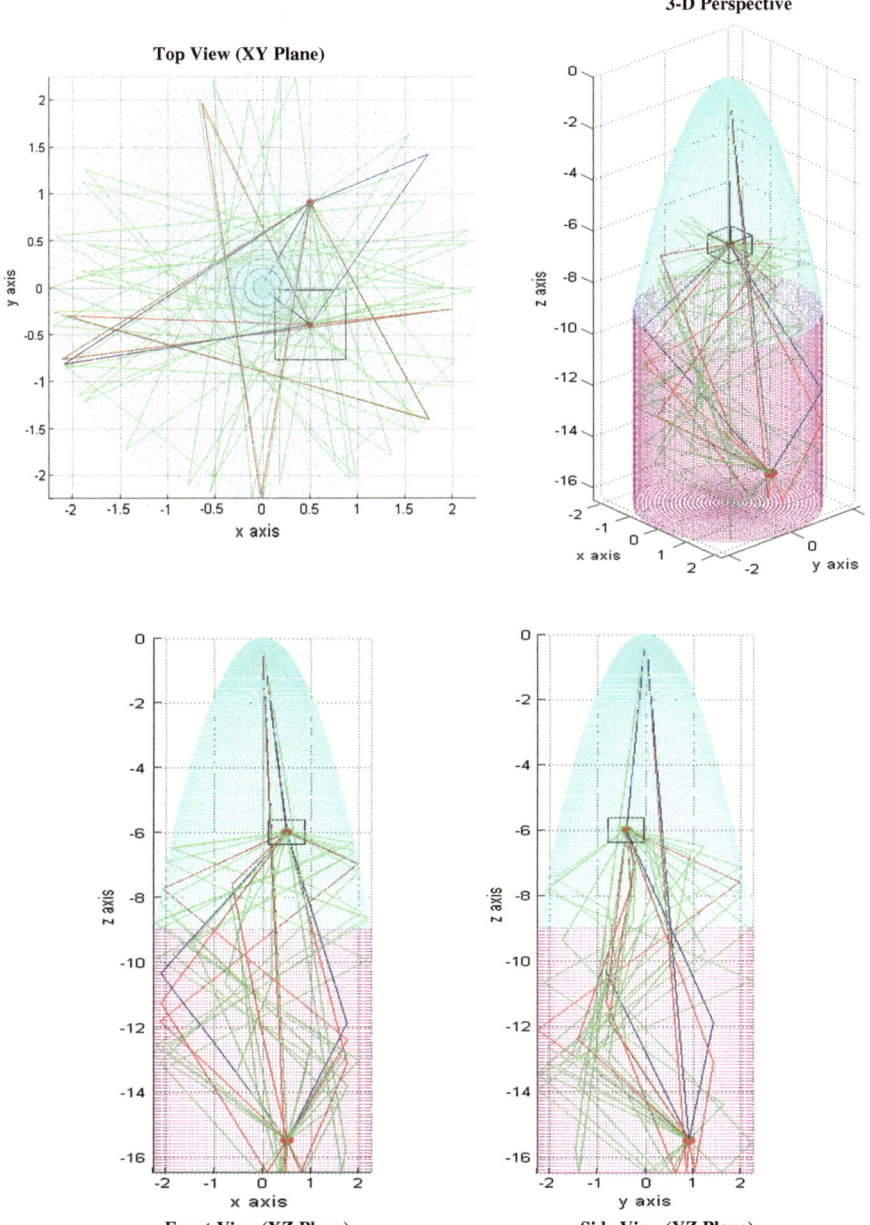

Fig. 38 *Cumulative rays* that reach the receiver up to *three bounces* visualized in *Matlab* (*Red dot* represents the source point and *green dot* represents the center of the reception sub-cube)

Table 11 Rays reaching receiver cumulatively up to N-bounce and the execution time (4,126,183 rays are launched)

N-bounce	No. of rays received	Program execution time
1 bounce	3	48 s
2 bounce	9	2.02 min
3 bounce	27	4.56 min
5 bounce	110	13.09 min
6 bounce	195	18.30 min
8 bounce	641	34.23 min
9 bounce	954	42.01 min
10 bounce	1269	47.30 min
15 bounce	1988	1.1 h
20 bounce	2088	1.11 h
25 bounce	2116	1.16 h
30 bounce	2164	1.23 h
35 bounce	2214	1.30 h

Table 12 The minimum and maximum time taken by the individual ray w.r.t. N-bounce

Bounce	T_{min} (µs)	T_{max} (µs)
1b	0.0606797832	0.1272235783
2b	0.0570701648	0.1338647486
3b	0.0628948028	0.1979992451
5b	0.0643434171	0.2003713978
6b	0.0753829836	0.2048255438
8b	0.0804982484	0.2069321281
9b	0.0855295197	0.2071127903
10b	0.0861130248	0.1680990475
15b	0.1446477235	0.2381960911
20b	0.2068371757	0.2664118835
25b	0.3153098920	0.3592535953
30b	0.3499112849	0.4272563892
1b	0.0606797832	0.1272235783
2b	0.0570701648	0.1338647486

The surface normal of the general paraboloid of revolution at a given point is given by Jha and Wiesbeck (1995)

$$x_n = \frac{2u \cos \phi}{\sqrt{a^2 + 4u^2}} \tag{7.d}$$

$$y_n = \frac{2u \sin \phi}{\sqrt{a^2 + 4u^2}} \tag{7.e}$$

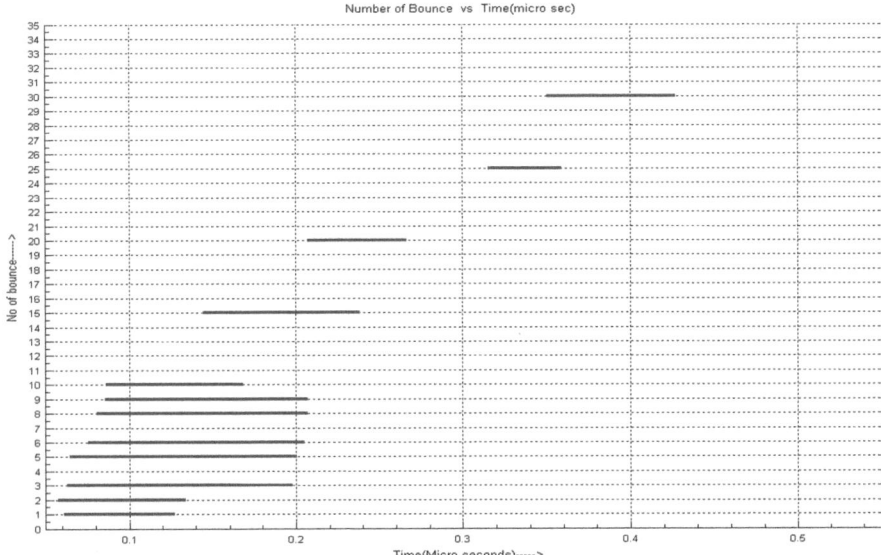

Fig. 39 T_{min}–T_{max} plot w.r.t. number of bounces for hybrid of GPOR and right circular cylinder

$$z_n = \frac{a}{\sqrt{a^2 + 4u^2}} \qquad (7.\text{f})$$

where a and u are the shaping parameters of the GPOR. The ray path details are calculated using the refined ray tracing algorithm as explained in Sect. 3.

Normal to the hybrid structure of GPOR–Right circular cylinder (Fig. 40) is visualized by the surface normal equation of the right circular cylinder and GPOR described above.

4.5.1   Results and Visualization of Ray Path

The ray path propagation inside the manned space module is visualized using *Matlab*. Table 13 considers 4,126,183 launched rays, out of which only 42 rays reach the receiver cumulatively up to three bounces excluding the direct ray. There are four rays, which reach after one bounce, 10 rays after two bounces, and 28 rays after three bounces. Figures 41, 42, and 43 give the visualization of one-bounce, two-bounce, and three-bounce rays reaching receiver. Figure 44 gives the cumulative number of rays till third bounce.

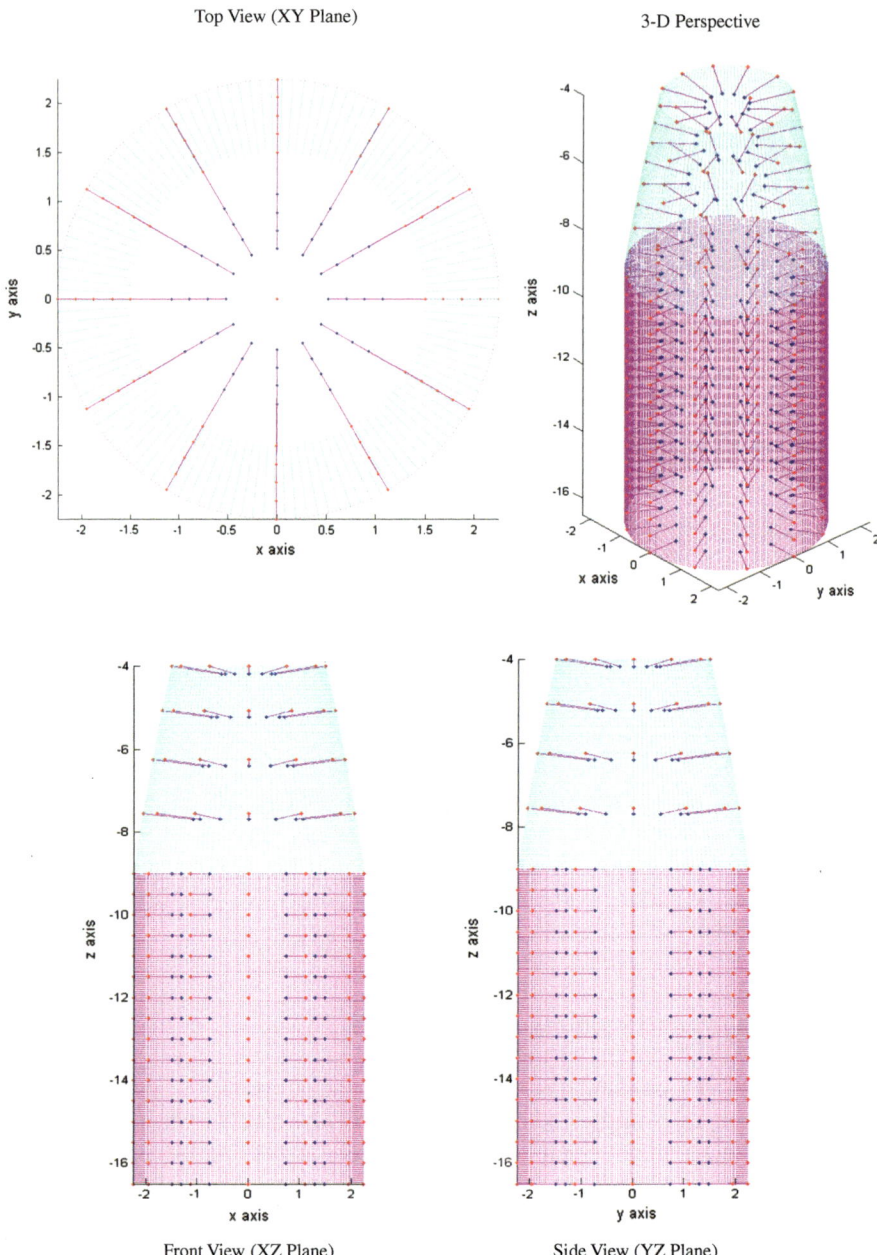

Fig. 40 Surface normal of hybrid structure (GPOR frustum and RCC) at different incident points

Table 13 Description of 42 rays that reached the receiver (till 3 bounce) out of 4,126,183 rays launched at 0.1° angular separation

Sl. no	θ	ϕ	Ex	Ey	Ez	Path length (m)	Time (μs)
1	40.386	212.569	1.485	0.164	−0.988	17.1297	0.057
2	40.450	298.406	−2.566	0.097	−1.753	17.796	0.059
3	31.151	213.504	−2.595	−0.418	−4.349	18.204	0.060
4	43.323	260.601	−1.137	2.364	−1.598	18.332	0.061
5	20.515	22.613	1.246	1.819	−5.892	18.570	0.061
6	45.0176	307.361	−1.016	−1.625	−1.769	18.868	0.062
7	144.801	212.519	1.501	0.164	−1.219	18.986	0.063
8	46.199	205.747	0.278	−1.522	−1.491	19.071	0.063
9	46.291	279.404	1.395	0.377	−0.343	19.628	0.065
10	45.766	294.563	1.007	1.589	−0.502	19.654	0.065
...							
...							
...							
31	159.671	213.501	−0.731	−0.117	2	23.755	0.079
32	173.593	316.529	0.899	0.246	1.995	24.064	0.080
33	173.550	270.002	−5.5E−05	1.186	−10.5	24.152	0.080
34	14.753	22.613	0.297	0.434	1.999	24.312	0.081
35	34.558	166.477	−0.299	−0.643	2.000	24.701	0.082
36	5.500	269.994	1.74E−05	0.192	2	26.120	0.087
37	167.184	22.534	0.258	0.376	2.006	27.596	0.091
38	175.206	269.994	1.62E−05	0.167	2.000	28.098	0.093
39	2.1518	269.939	0.000	0.394	−10.500	47.033	0.156
40	9.450	213.529	−1.725	−0.279	−10.5	47.646	0.158
41	178.005	270.058	−0.000	0.365	−10.501	49.030	0.163
42	1.933	270.000	−3.3E−07	0.067	2	51.029	0.170

4.5.2   Discussion

The number of rays that reach the receiver w.r.t. N-bounce (cumulatively) and the program execution time of the corresponding bounce is given in Table 14. The minimum and maximum time taken by the rays to reach the receiver w.r.t. N-bounce is given in Table 15, and Figure 39 gives the time plot of the same (Figure 45).

Top View (XY Plane)   3-D Perspective

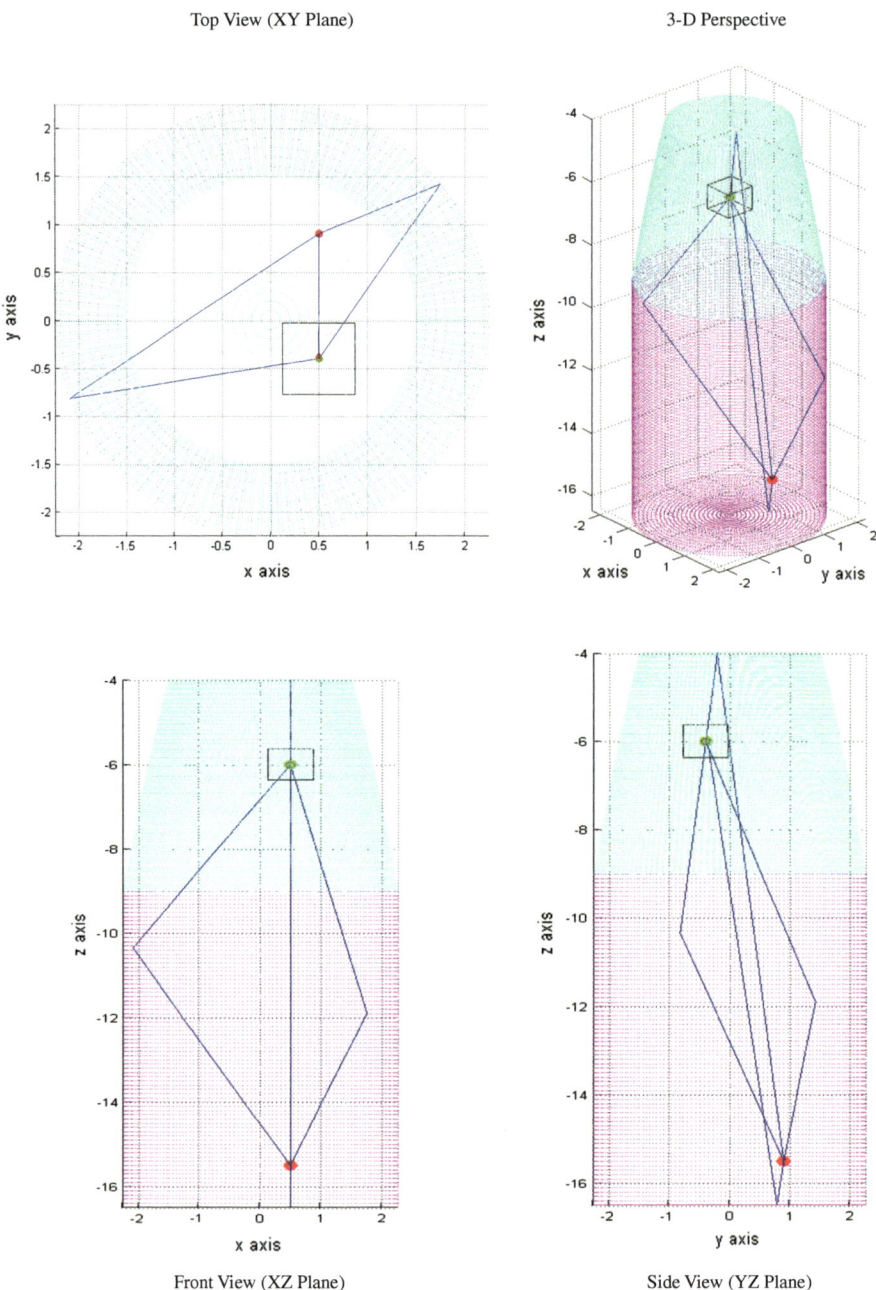

Front View (XZ Plane)   Side View (YZ Plane)

Fig. 41 Ray path of rays that reaches receiver after *single bounce* (*Red dot* represents the source point and *green dot* represents the center of the reception sub-cube)

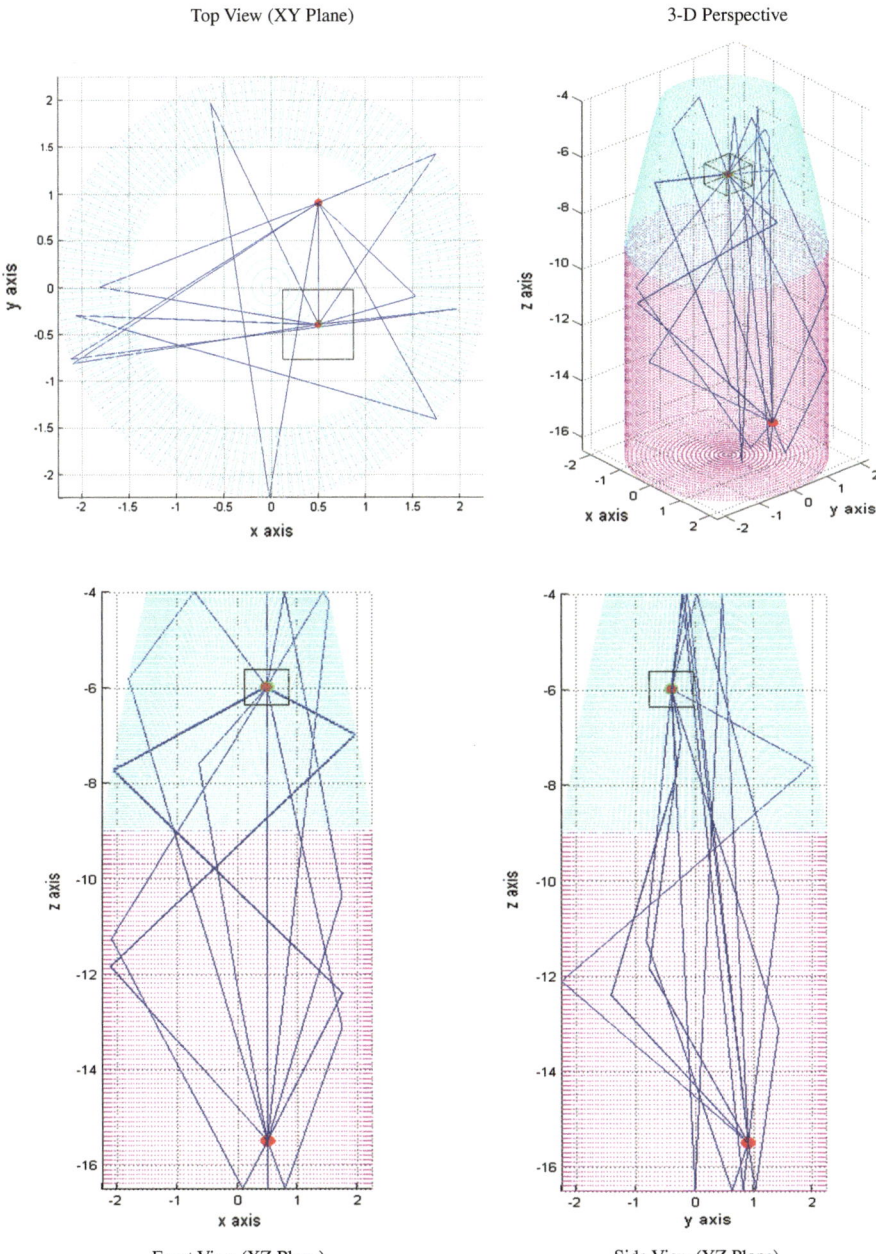

Fig. 42 Ray path of rays that reach receiver after *two bounces* (*Red dot* represents the source and center of the reception sub-cube represents the receiver)

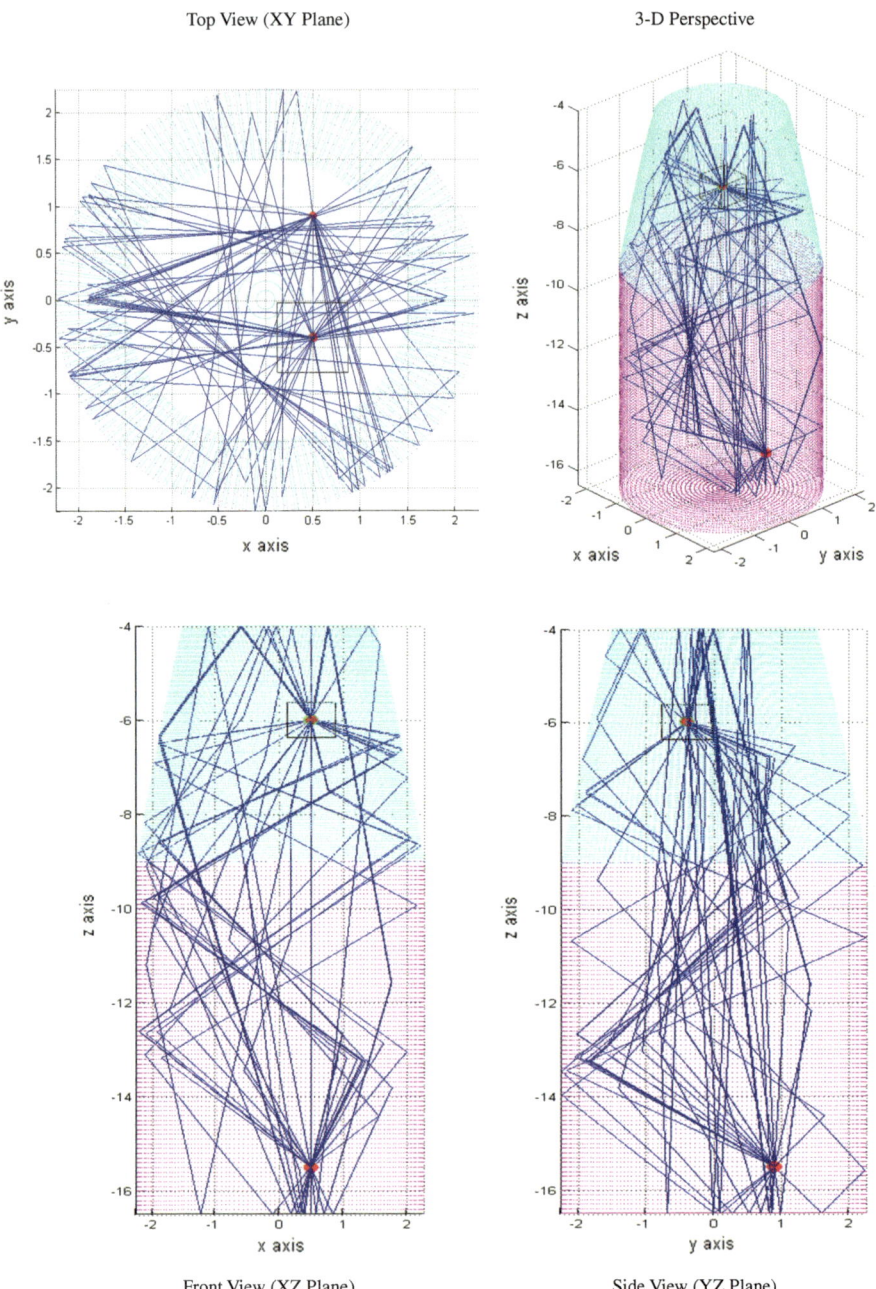

Fig. 43 Ray path of rays that reach receiver after *three bounces* (*Red dot* represents the source and center of the reception sub-cube represents the receiver)

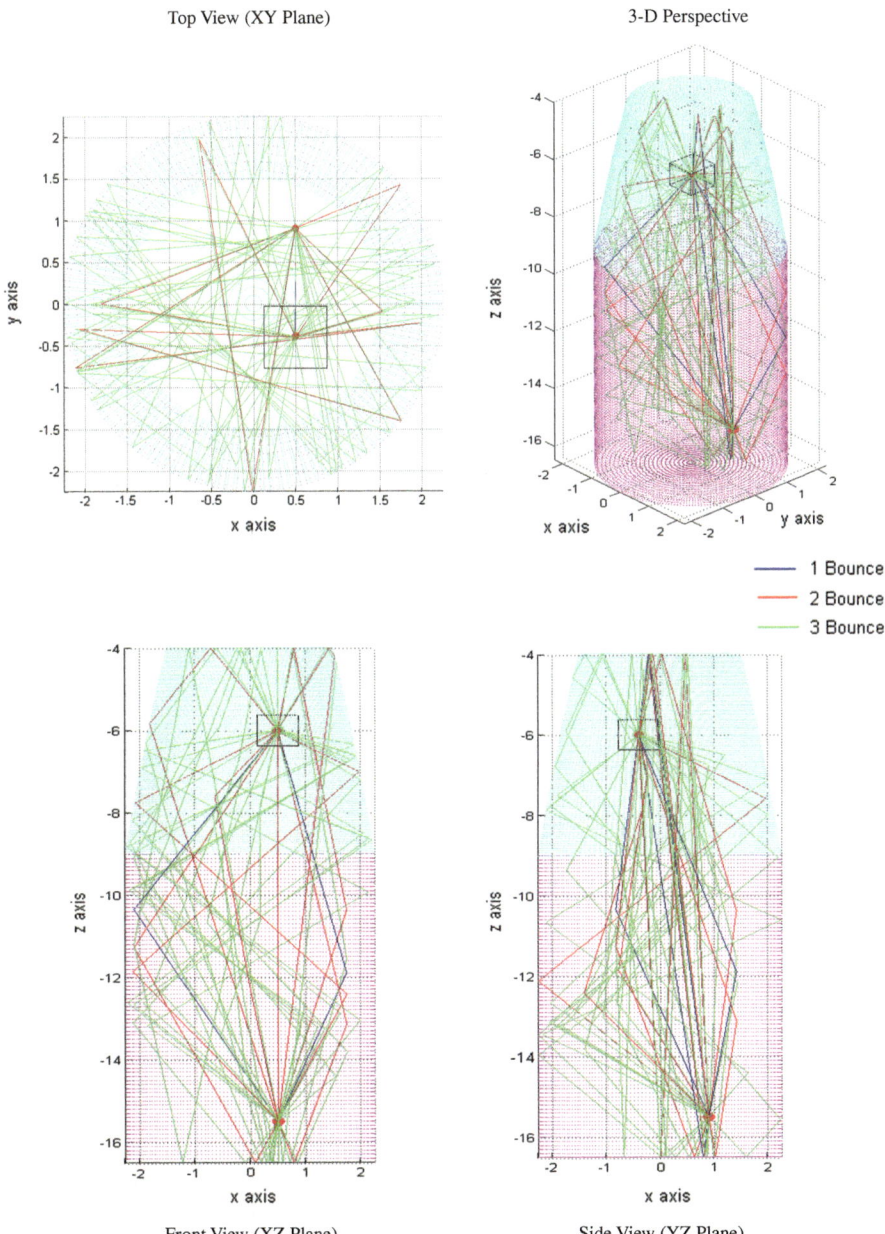

Fig. 44 *Cumulative rays* that reach receiver up to *three bounces* visualized in *Matlab* (*Red dot* represents the source and center of the reception sub-cube represents the receiver)

Table 14 Rays reaching receiver cumulatively up to N-bounce and the execution time (4,126,183 rays are launched)

N-bounce	No. of rays received	Program execution time
1 bounce	4	25 S
2 bounce	14	35 S
3 bounce	42	54 S
5 bounce	120	1.55 min
6 bounce	187	2.14 min
8 bounce	370	4.05 min
9 bounce	450	5.55 min
10 bounce	538	6.67 min
15 bounce	893	11 min
20 bounce	1082	17.56 min
25 bounce	1205	24.35 min
30 bounce	1279	29.45 min
35 bounce	1318	34.12 min
40 bounce	1345	42.35 min

Table 15 The minimum and maximum time taken by the individual ray w.r.t. N-bounce

Bounce	T_{min} (µs)	T_{max} (µs)
1b	0.0606819	0.0870675
2b	0.0570990	0.1567776
3b	0.0628949	0.1700968
5b	0.0743642	0.2533867
6b	0.0798357	0.2466307
8b	0.0901950	0.2628089
9b	0.1008552	0.3283111
10b	0.1113745	0.3396298
15b	0.1844066	0.5065902
20b	0.2935481	0.6012170
25b	0.3116375	0.7544882
30b	0.4006642	0.9196970
35b	0.9452857	1.021048
40b	1.082698	1.187521

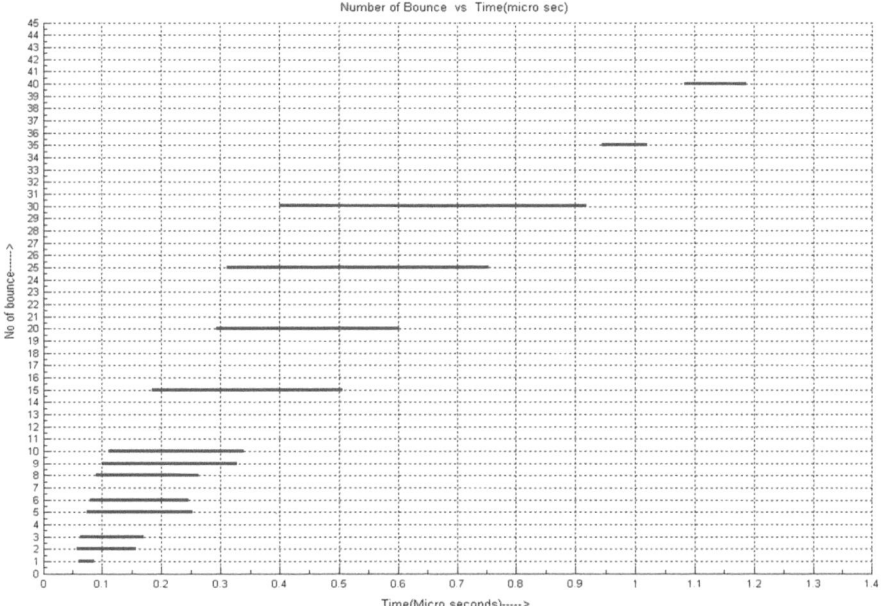

Fig. 45 T_{min}–T_{max} plot w.r.t. number of bounces for space module

5 Conclusion

In this document, effect of shaping parameter for single- and double-curvatured surfaces in geometric ray tracing is reported. As most of the complex aerospace structures can be modeled as hybrid of QUACYLs and QUASORs, the analysis of geometric ray tracing inside is the basic requirement to analyze the RF build-up. Toward this, ray tracing inside different quadric surfaces such as right circular cylinder, GPOR, and GPOR frustum of different shaping parameters are carried out and the ray path details are visualized. Finally, ray tracing inside a space module, which is a hybrid of a finite segment of right circular cylinder and a frustum of *general paraboloid of revolution* (GPOR), is reported.

The ray path data inside the single- and double-curvatured structures is generated using a refined ray tracing algorithm, which involves an *uniform ray launching* scheme, an intelligent scheme for *ray bunching*, and an *adaptive reception* procedure. The minimum and maximum path lengths for individual bounces (i.e., for one bounce, two bounces, etc.) are also reported, which is a required parameter for RF analysis inside the hybrid structures.

Like planar structures, inside concave structures, the T_{min}–T_{max} plot is not linear as there is always a solution, which takes a helical path and reaches the receiver covering the smallest path length of the corresponding bounces. This is always the case for higher bounces.

As the curvature effect is more, the rays reaching receiver become insignificant because they undergo successive reflections, similar to surface wave diffraction. This effect leads to attenuation of the EM wave and the field will converge with lower number of bounce.

References

Choudhury, B., S. Hema, J.P. Bomer and R.M. Jha. 2013. RF field mapping inside large passenger aircraft cabin using refined ray tracing algorithm. *IEEE Antennas and Propagation Magazine* 55(1): 276–288.

Jackson, P. (ed.), Jane's All the World's Aircraft 2007–2008. Jane's Information Group, 973 p., Coulsdon, (2007–2008). ISBN: 13-978-07106-2792-6.

Jha, R.M., and W. Wiesbeck. 1995. Geodesic Constant Method: A novel approach to analytical surface-ray tracing on convex conducting bodies. *IEEE Antennas and Propagation Magazine* 37: 28–38.

Kreyszig, E. 2010. *Advanced engineering mathematics*, 1280 p., 10th edn. Hoboken: Wiley. ISBN: 978-04-704-5836-5.

Moon, P., and D.E. Spencer. (1971). *Field theory handbook*. 2nd edn. Heidelberg: Springer. ISBN: 3-540-02732-7.

Seidel, S.Y., and T.S. Rappaport. 1994. Site-specific propagation prediction for wireless in building personal communication system design. *IEEE Transactions on Vehicular Technology* 43: 879–891.

About the Book

This book describes the ray tracing effects inside different quadric surfaces. Analytical surface modeling is a priori requirement for electromagnetic (EM) analysis over aerospace platforms. Although numerically specified surfaces and even non-uniform rational basis spline (NURBS) can be used for modeling such surfaces, for most practical EM applications, it is sufficient to model them as quadric surface patches and the hybrids thereof. It is therefore apparent that a vast majority of aerospace bodies can be conveniently modeled as combinations of simpler quadric surfaces, i.e., hybrid of quadric cylinders and quadric surfaces of revolutions. Hence, the analysis of geometric ray tracing inside is prerequisite to analyze the RF build-up. This book describes the ray tracing effects inside different quadric surfaces such as right circular cylinder, general paraboloid of revolution (GPOR), GPOR frustum of different shaping parameters, and the corresponding visualization of the ray-path details. Finally, ray tracing inside a typical space module, which is a hybrid of a finite segment of right circular cylinder and a frustum of GPOR, is analyzed for practical aerospace applications.

© The Author(s) 2016
B. Choudhury and R.M. Jha, *Refined Ray Tracing Inside Single-
and Double-Curvatured Concave Surfaces*, SpringerBriefs
in Computational Electromagnetics, DOI 10.1007/978-981-287-808-3

Author Index

© The Author(s) 2016
B. Choudhury and R.M. Jha, *Refined Ray Tracing Inside Single-and Double-Curvatured Concave Surfaces*, SpringerBriefs
in Computational Electromagnetics, DOI 10.1007/978-981-287-808-3

Subject Index

© The Author(s) 2016
B. Choudhury and R.M. Jha, *Refined Ray Tracing Inside Single-
and Double-Curvatured Concave Surfaces*, SpringerBriefs
in Computational Electromagnetics, DOI 10.1007/978-981-287-808-3